UNDERSTANDING
VECTOR
CALCULUS

UNDERSTANDING
VECTOR
CALCULUS

Practical Development and Solved Problems

JERROLD FRANKLIN

Temple University

DOVER PUBLICATIONS, INC.
MINEOLA, NEW YORK

Bibliographical Note

Understanding Vector Calculus: Practical Development and Solved Problems
is a new work, first published by Dover Publications, Inc., in 2020.

International Standard Book Number
ISBN-13: 978-0-486-83590-7
ISBN-10: 0-486-83590-1

Manufactured in the United States by LSC Communications
83590101
www.doverpublications.com

2 4 6 8 10 9 7 5 3

2020

Contents

Preface

My purpose in this development of vector calculus is to present it in a way that will prove useful to you for the rest of your career in science or mathematics. You can think of this presentation as a 'workbook' for learning how to use vector calculus in practical calculations and derivations. After studying the text and doing the problems, you should not have to memorize long equations or need to look anything up while you work in your field. I want you to get to the point where you can treat the use of vector calculus on the same level as you would simple algebraic calculations, working them out as you go.

The necessary background for using this book is just a knowledge of one-dimensional differential and integral calculus. Having had a multivariable calculus course is unnecessary because the multivariable aspect of vector calculus will be developed from scratch. In fact, you might have to unlearn some of the things covered in a multivariable calculus course if it relied too much on using coordinate systems, rather than the vector methods we will introduce.

Although the examples in the book are usually related to physical examples, particularly electromagnetism, it is not necessary to have any background in physics. If any of the physics is new to you, you can simply disregard it and concentrate on the mathematics. On the other hand, if you have a background in (relatively simple) physics, I think you will find the references to physics of interest. In treating electromagnetism, we use what are sometimes called 'natural units' in which the potential of a point charge q is given by q/r. In this way, we avoid the introduction of artificial constants that might complicate the mathematics.

The book has two parts. Part one is a brief text developing vector calculus from the very beginning, and then including some more detailed applications. Part two consists of answered problems, which are all closely related to the development of vector calculus in the text. Although there are answers immediately following each problem, you should not be too quick to use the answer as a crutch. For problems that involve working through a calculation, please try your hardest to do it on your own. Then you can use the answer to check your result, or possibly see another way of doing the problem. If you go immediately to the answer, you will be learning about vector calculus, but not how to use vector calculus. Of course, if you don't see how to start, or run into a roadblock, the answer is there to help you.

In any event, I hope you find my treatment interesting, perhaps entertaining, and above all, useful. I think you will see that you have nothing to fear from vector calculus, and that it will prove to be a good friend for the rest of your life. Feel free to use me as a resource, either by answering questions you might have about the material in the book, or about anything else. You can contact me at Jerry.F@Temple.edu.

UNDERSTANDING
VECTOR
CALCULUS

Chapter 1

Vector Differential Operators

1.1 Gradient

A scalar field (that is, a scalar function of the position vector $\mathbf{r}$) is conveniently pictured by means of surfaces (in three dimensions) or lines (in two dimensions) along which its magnitude is constant. Depending on the physical application, these constant magnitude surfaces or lines could be called equipotentials, isobars, isotherms, or whatever applies in the given situation.

A common example, shown in Fig. 1.1, is a topographic map of a hillside.

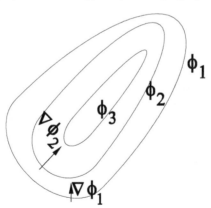

Figure 1.1: Equipotentials ϕ_i and gradients $\nabla \phi_i$.

The lines of equal altitude shown on the map are equipotentials of the gravitational field. No work is done in moving along an equipotential, and the direction of steepest slope is everywhere perpendicular to the equipotential. That perpendicular direction is defined as the direction of the **gradient** of the potential. In our topographic example, this direction is the steepest direction up the hill.

1

Experimentally, this would be opposite to the direction a ball would roll if placed at rest on the hillside.

The magnitude of the gradient is defined to be the rate of change of the potential with respect to distance in the direction of maximum increase. This provides a mathematical definition of the gradient as

$$\mathbf{grad}\phi = \hat{\mathbf{n}}\frac{d\phi}{|\mathbf{dr}|}. \tag{1.1}$$

In Eq. (1.1), the unit vector $\hat{\mathbf{n}}$ ($=\mathbf{n}/|\mathbf{n}|$) is in the direction of maximum increase of ϕ, and $\mathbf{dr}$ is taken in that direction of maximum increase.

For infinitesimal displacements, an equipotential surface can be approximated by its tangent plane (or tangent line in two dimensions), so the change in a scalar field in an infinitesimal displacement $\mathbf{dr}$ will vary as the cosine of the angle between the direction of maximum gradient and $\mathbf{dr}$. Then the differential change of ϕ in any direction is given by

$$d\phi(\mathbf{r}) = \mathbf{dr}\cdot\mathbf{grad}\phi. \tag{1.2}$$

This relation can be considered an alternate definition of the gradient.

The rate of change of a scalar field in a general direction $\hat{\mathbf{a}}$, not necessarily the direction of maximum change, can be defined by choosing $\mathbf{dr}$ in that direction and dividing both sides of Eq. (1.2) by the magnitude $|\mathbf{dr}|$. This is called the **directional derivative** of ϕ, defined by $\hat{\mathbf{a}}\cdot\mathbf{grad}\phi$ for the rate of change of ϕ in the direction $\hat{\mathbf{a}}$.

We see that $\mathbf{grad}\phi$ is a vector derivative of the scalar function ϕ. However, the operation by $\mathbf{grad}$ has more general applications. For this reason, we introduce it as a **vector differential operator**, which we denote as ∇ (pronounced 'del'). The actual calculation of $\nabla\phi$ can be made using a coordinate system, but it is usually better to use the definition of the gradient in Eq. (1.1) or (1.2) to find it for various functions of the position vector $\mathbf{r}$ directly.

We start with r ($= |\mathbf{r}|$), the magnitude of $\mathbf{r}$, treated as a scalar. Its maximum rate of change is in the $\hat{\mathbf{r}}$ direction, and its derivative in that direction is $dr/dr = 1$, so

$$\nabla r = \hat{\mathbf{r}}. \tag{1.3}$$

Next, we consider any scalar function, $f(r)$, of the magnitude of $\mathbf{r}$. The direction of maximum rate of change of $f(r)$ will also be $\hat{\mathbf{r}}$, and its derivative in that direction is df/dr, so

$$\nabla f(r) = \hat{\mathbf{r}}\frac{df}{dr}. \tag{1.4}$$

In electrostatics, the electric field is the negative gradient of the potential. A vector field that is the gradient of a scalar field is said to be **derivable from**

a potential. As an example, applying this for the Coulomb's law potential due to a point charge ($\phi = q/r$), the electric field is given by

$$\mathbf{E} = -\boldsymbol{\nabla}\phi = -\boldsymbol{\nabla}\left(\frac{q}{r}\right) = -q\hat{\mathbf{r}}\frac{d}{dr}\left(\frac{1}{r}\right) = q\frac{\hat{\mathbf{r}}}{r^2}. \tag{1.5}$$

Equation (1.5) derives the electric field from the potential. We can also derive the potential from the electric field by integrating Eq. (1.2) to give

$$\phi(\mathbf{r}) = \int_{\infty}^{\mathbf{r}} d\phi = \int_{\infty}^{\mathbf{r}} \mathbf{dr}\cdot\boldsymbol{\nabla}\phi = \int_{\mathbf{r}}^{\infty} \mathbf{dr}\cdot\mathbf{E} \tag{1.6}$$

In Eq. (1.6), we have chosen the boundary condition that the potential ϕ approaches zero as r goes to infinity.

An interesting, and useful, property of the gradient is that its line integral around a closed path equals zero. That is

$$\oint \mathbf{dr}\cdot\boldsymbol{\nabla}\phi = \oint d\phi = 0. \tag{1.7}$$

(The notation $\oint$ means the line integral is around a closed path.) Since the electric field is the negative gradient of the potential ($\mathbf{E} = -\boldsymbol{\nabla}\phi$), its integral around a closed path is zero. This means that $\mathbf{E}$ is a **conservative field** that does no work moving a charge around a closed path. This result applies to any vector field (for instance, the gravitational field) that is the gradient of a scalar field.

We can also apply the gradient to an integral, for instance taking the negative gradient of the potential for a charge distribution ρ (charge per unit volume) to get the corresponding electric field. The generalization of Coulomb's law for the potential due to a charge distribution is given by

$$\phi(\mathbf{r}) = \int \frac{\rho(\mathbf{r}')d^3r'}{|\mathbf{r} - \mathbf{r}'|}. \tag{1.8}$$

The corresponding electric field is given by the negative gradient of this integral:

$$\begin{aligned}
\mathbf{E}(\mathbf{r}) &= -\boldsymbol{\nabla}\int \frac{\rho(\mathbf{r}')d^3r'}{|\mathbf{r} - \mathbf{r}'|} \\
&= -\int \rho(\mathbf{r}')d^3r'\boldsymbol{\nabla}\left[\frac{1}{|\mathbf{r} - \mathbf{r}'|}\right] \\
&= \int \frac{(\mathbf{r} - \mathbf{r}')\rho(\mathbf{r}')d^3r'}{|\mathbf{r} - \mathbf{r}'|^3}.
\end{aligned} \tag{1.9}$$

We could take the $\boldsymbol{\nabla}$ inside the integral because it acts only on $\mathbf{r}$ and not on the integration variable $\mathbf{r}'$. In taking the gradient of $1/|\mathbf{r} - \mathbf{r}'|$, we have used the fact that the introduction of the constant vector $\mathbf{r}'$ just shifts the origin from $\mathbf{0}$ to the position $\mathbf{r}'$. That is,

$$\nabla\left(\frac{-1}{r}\right) = \frac{\mathbf{r}}{r^3} \Longrightarrow \nabla\left(\frac{1}{|\mathbf{r}-\mathbf{r}'|}\right) = \frac{-(\mathbf{r}-\mathbf{r}')}{|\mathbf{r}-\mathbf{r}'|^3}. \tag{1.10}$$

We can give an explicit form for the operator ∇ in Cartesian (x, y, z) coordinates, where an infinitesimal displacement is given by

$$\mathbf{dr} = \hat{\mathbf{i}}dx + \hat{\mathbf{j}}dy + \hat{\mathbf{k}}dz. \tag{1.11}$$

Then, Eq. (1.2) defining the gradient can be written as

$$\begin{aligned}
d\phi(x, y, z) &= \mathbf{dr} \cdot \nabla\phi \\
&= (\hat{\mathbf{i}}dx + \hat{\mathbf{j}}dy + \hat{\mathbf{k}}dz) \cdot \nabla\phi \\
&= (\nabla\phi)_x dx + (\nabla\phi)_y dy + (\nabla\phi)_z dz.
\end{aligned} \tag{1.12}$$

At the same time, the differential of the function $\phi(x, y, z)$ of three variables is given by

$$d\phi(x, y, z) = (\partial_x\phi)dx + (\partial_y\phi)dy + (\partial_z\phi)dz. \tag{1.13}$$

($\partial_x\phi$, the **partial derivative** of ϕ, differentiates ϕ with respect to the variable x keeping the other variables fixed. We will use the notation ∂_x rather than the more cumbersome $\frac{\partial}{\partial x}$ to represent the partial derivative.)

Comparing Eqs. (1.12) and (1.13) for the same differential, and using the fact that the displacements dx, dy, dz are independent and arbitrary, we see that

$$\nabla\phi = \hat{\mathbf{i}}(\partial_x\phi) + \hat{\mathbf{j}}(\partial_y\phi) + \hat{\mathbf{k}}(\partial_z\phi). \tag{1.14}$$

Equation (1.14) shows that the vector differential operator ∇ is given in Cartesian coordinates by

$$\nabla = \hat{\mathbf{i}}\partial_x + \hat{\mathbf{j}}\partial_y + \hat{\mathbf{k}}\partial_z. \tag{1.15}$$

Equation (1.15) holds only in Cartesian coordinates. The explicit expansion of ∇ given by Eq. (1.15) is more complicated in other coordinate systems. The derivation of those expansions will be given in a later chapter, and we will derive a coordinate independent form of ∇ later in the text. Further properties and uses of the gradient will be given as answers to the gradient problems.

1.2   Divergence

A vector field can have two different types of variation. It can vary along its direction, for instance like the velocity field, $\mathbf{v}$, of a stream as the slope gets steeper. The vector field can also vary across its direction, as when the velocity is faster in the middle of the stream than near the edges. How can these two variations be measured?

The rate of increase of a vector field along its direction is called the **divergence** of the vector field. A simple example for a vector field **E** is shown in Fig. 1.2. A measure of the strength of the field is the density of **lines of force** in the figure, with the increase in the field indicated by increasing lines of force.

We construct a mathematical volume V enclosed by a surface S, as shown in the figure. The increase in **E** can be seen in the figure as more lines of **E** leaving the volume than entering it ('diverging' from the volume).

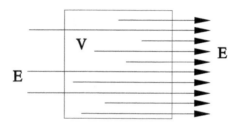

Figure 1.2: Divergence of lines of E. More lines leave the volume than enter it.

A quantitative measure of the excess of lines leaving the volume is given by the integral $\oint_S d\mathbf{A} \cdot \mathbf{E}$.[1] This integral can be used to define an average divergence (written as 'div') of the lines of the vector field. That is

$$\langle \text{div}\,\mathbf{E} \rangle_V = \frac{1}{V} \oint_S d\mathbf{A}' \cdot \mathbf{E}(\mathbf{r}'), \tag{1.16}$$

where the notation $\langle \text{div}\,\mathbf{E} \rangle_V$ denotes the average of div **E** over the volume V.

The value of div **E** at a point **r** can be defined by shrinking the integral about the point, so

$$\text{div}\,\mathbf{E}(\mathbf{r}) = \lim_{V \to 0} \frac{1}{V} \oint_S d\mathbf{A}' \cdot \mathbf{E}(\mathbf{r}') \tag{1.17}$$

---

[1]The notation $\oint_S$ means that the integral is over a closed surface. The vector differential area $d\mathbf{A}$ is an infinitesimal area that can be approximated by a plane, with its vector direction being perpendicular to the plane. By convention, the positive direction of the vector $d\mathbf{A}$ is out of the closed surface. This means that the integral $\oint_S d\mathbf{A} \cdot \mathbf{E}$ is an integral of the outward normal component of the vector **E** over a closed surface. Using an explicit coordinate system (such as Cartesian coordinates) can make the integral more complicated.

gives the divergence of the vector field at the point $\mathbf{r}$ (if the limit exists), and is a measure of its rate of increase along the direction of the vector field. We take Eq. (1.17) as the definition of the divergence operator.

We will show on the next page that div $\mathbf{E}$ can be written as $\boldsymbol{\nabla}\!\cdot\!\mathbf{E}$, corresponding to the dot product of the vector differential operator $\boldsymbol{\nabla}$ with a vector $\mathbf{E}$. We start using that notation now, so that the following equations will be in the usual notation for the divergence.

1.2.1   Divergence Theorem

Combining Eq. (1.16) for the average value of the divergence over a finite volume V with the definition of a volume average that

$$\langle \boldsymbol{\nabla}\!\cdot\!\mathbf{E}\rangle_V = \frac{1}{V}\int_V \boldsymbol{\nabla}\!\cdot\!\mathbf{E}d^3r, \tag{1.18}$$

we get

$$\int_V \boldsymbol{\nabla}\!\cdot\!\mathbf{E}d^3r = \oint_S d\mathbf{A}\!\cdot\!\mathbf{E}. \tag{1.19}$$

In this form, it is called the **divergence theorem**.

The definition of the divergence given by Eq. (1.17) can be used to evaluate the divergence of the position vector. We apply the definition to a sphere of radius R, getting[2]

$$
\begin{aligned}
\boldsymbol{\nabla}\!\cdot\!\mathbf{r} &= \lim_{V\to 0}\frac{1}{V}\oint_S d\mathbf{A}'\!\cdot\!\mathbf{r}' = \lim_{R\to 0}\frac{1}{V}\oint R^3 d\Omega' \\
&= \lim_{R\to 0}\frac{4\pi R^3}{(4/3)\pi R^3} = 3.
\end{aligned}
\tag{1.20}
$$

As we did with the gradient, we now show what the divergence would look like in Cartesian coordinates. Figure 1.3 shows an infinitesimal volume (a parallelepiped in Cartesian coordinates) of dimensions $\Delta x \times \Delta y \times \Delta z$, that will shrink to zero at the point x, y, z. The surface integral in the definition of div$\mathbf{E}$ is over the six faces of the parallelepiped, I-VI, so the integral can be written as

$$\lim_{V\to 0}\frac{1}{V}\oint_S d\mathbf{A}\!\cdot\!\mathbf{r} = I + II + III + IV + V + VI, \tag{1.21}$$

where I indicates the integral over face I, and similarly for the other faces.

---

[2]The differential of a **solid angle** in this derivation is defined by $d\Omega = \hat{\mathbf{r}}\!\cdot\!d\mathbf{A}/r^2$. Although a solid angle is intrinsically dimensionless, it is given the name **steradian**, in analogy with the name radian given for the size of an angle.

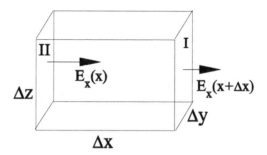

Figure 1.3: Volume element in Cartesian coordinates.

We concentrate first on faces I and II, each parallel to the y-z plane. In the limit as both Δy and Δz approach zero, the integral over face I approaches $E_x(x+\Delta x, y, z)\Delta y \Delta z$, and that over face II approaches $-E_x(x, y, z)\Delta y \Delta z$, provided that E_x is continuous at the point x, y, z. Then, for these two faces

$$
\begin{aligned}
I + II &= \lim_{\Delta x, \Delta y, \Delta z \to 0} \frac{[E_x(x + \Delta x, y, z)\Delta y \Delta z - E_x(x, y, z)\Delta y \Delta z]}{\Delta x \Delta y \Delta z} \\
&= \lim_{\Delta x \to 0} \frac{[E_x(x + \Delta x, y, z) - E_x(x, y, z)]}{\Delta x} \\
&= \partial_x E_x, && (1.22)
\end{aligned}
$$

where the last step follows from the definition of the partial derivative.

The integrals over the other four faces are done in the same way, leading to similar results, with the substitutions $x \to y$ and then $x \to z$, so

$$
\text{div}\,\mathbf{E} = \partial_x E_x + \partial_y E_y + \partial_z E_z \qquad (1.23)
$$

in Cartesian coordinates. The form of this equation suggests that $\text{div}\,\mathbf{E}$ could be written as a dot product

$$
\text{div}\,\mathbf{E} = \boldsymbol{\nabla} \cdot \mathbf{E}, \qquad (1.24)
$$

of the vector differential operator $\boldsymbol{\nabla}$ with $\mathbf{E}$. This can be stated in words as "Divergence E equals Del dot E." The representation of div by the vector differential operator $\boldsymbol{\nabla}\cdot$ is not limited to Cartesian coordinates, although the explicit form given in Eq. (1.23) does not hold for other coordinate systems.

1.3 Curl

Next, we look at how a vector $\mathbf{E}$ can vary across its direction, and we give a physical definition of the **curl** of a vector field. Figure 1.4 shows a vector field having such a variation, with the density of lines being proportional to the strength of the field. If this were a velocity field, such as the current of water in a stream, this variation could be measured experimentally by placing a paddle wheel in the stream as shown in the figure. Then the rotation of the paddle wheel would be a measure of the variation of the vector field. This can be done without getting wet by calculating a line integral around a typical closed curve C, as shown on the figure.

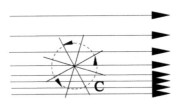

Figure 1.4: Velocity field with curl. The current increases going down on the figure causing the paddle wheel to rotate counterclockwise.

The line integral can be used to define an average value of the variation (called **curl**) over a surface S bounded by the curve C. The average **curl** is defined by

$$\langle \hat{\mathbf{n}} \cdot \mathbf{curl}\,\mathbf{E}\rangle_S = \frac{1}{S}\oint_C d\mathbf{r}' \cdot \mathbf{E}(\mathbf{r}'), \qquad (1.25)$$

where $\hat{\mathbf{n}}$ is the unit vector normal to the surface S at any point. Note that, by this definition, the vector average $\langle \hat{\mathbf{n}} \cdot \mathbf{curl}\,\mathbf{E}\rangle_S$ does not depend on the shape of the surface S, but only on the bounding path C and the area of the surface. Since the variation will be different in different directions, it is the average value of the normal component of **curl** that is defined by Eq. (1.25).

The positive sign for the direction of $\hat{\mathbf{n}}$ is taken by convention to be the **boreal** direction. That is, if the integral around the contour C is taken in the direction of the rotation of the earth, then the north pole is in the positive direction as shown on Fig. 1.5a.

This is also stated as the **right hand rule**: If the integral around the contour C is taken in the direction that the four fingers of the right hand curl as they tend to close, then the right thumb points in the positive direction for $\hat{\mathbf{n}}$, as shown in Fig. 1.5b. This will be our general sign convention relating the direction of integration around a closed curve and the positive direction of the normal vector to any surface bounded by the curve.

The value of **curl E** at a point $\mathbf{r}$ can be defined by starting with a smooth surface through the point and taking the limit as the curve bounding the surface shrinks about the point, and the enclosed surface shrinks to zero area. This gives

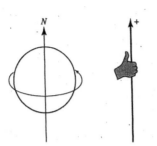

Figure 1.5: (a) **Boreal direction on the globe.** (b) **Right hand rule for positive direction.**

the definition of **curl** at a point:

$$[\mathbf{curl}\,\mathbf{E(r)}]_n = \lim_{S \to 0} \frac{1}{S} \oint_C d\mathbf{r}' \cdot \mathbf{E(r')}. \qquad (1.26)$$

As the curve C shrinks to a point, the smooth surface approaches its tangent plane at the point, and $(\mathbf{curl}\,\mathbf{E})_n$ in Eq.(1.26) represents the component of curl in the direction of the normal vector $\hat{\mathbf{n}}$ to the tangent plane.

We will show shortly that $\mathbf{curl}\,\mathbf{E}$ can be written as $\nabla \times \mathbf{E}$, corresponding to the cross product of the vector differential operator ∇ with a vector $\mathbf{E}$. We start using that notation now.

We can use the definition of **curl** in Eq. (1.26) to show that the **curl** of any gradient is zero:

$$\nabla \times \nabla \phi = 0. \qquad (1.27)$$

This is because any gradient is a conservative vector field and the closed contour integral on the right hand side of Eq. (1.26) vanishes. As a mnemonic, the fact that $\nabla \times \nabla \phi = 0$ can be thought of as following from the vanishing of the cross product of a vector with itself. But that is not strictly the case for vector differential operators, and an important counterexample is given as a problem.

Since the position vector $\mathbf{r}$ is a gradient ($\nabla r^2 = 2\mathbf{r}$), it follows that

$$\nabla \times \mathbf{r} = 0. \qquad (1.28)$$

Also, the curl of the electrostatic field $\mathbf{E}$ vanishes because it is the negative gradient of a potential. A vector field with zero curl is said to be **irrotational**.

With $\mathbf{curl}\,\mathbf{E}$ defined by Eq. (1.26), it is possible to find its specific form in Cartesian coordinates. To find $\mathbf{curl}\,\mathbf{E}$ at the point x, y, z, we consider an infinestimal rectangle of dimension Δx by Δy parallel to the x-y plane, as shown in Fig. 1.6.

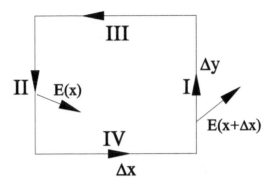

Figure 1.6: Differential surface for curl in Cartesian coordinates

The line integral around the rectangle consists of four parts, so

$$\lim_{S \to 0} \frac{1}{S} \oint_C \mathbf{dr} \cdot \mathbf{E} = I + II + II + IV. \tag{1.29}$$

As Δy approaches zero, we can make the replacement

$$\int_y^{y+\Delta y} f(y')dy' \to f(y)\Delta y, \tag{1.30}$$

provided that $f(y)$ is continuous at y. Then the contribution to Eq. (1.26) from sides I and II is

$$
\begin{aligned}
I + II &= \lim_{\Delta x, \Delta y \to 0} \frac{[E_y(x + \Delta x, y, z)\Delta y - E_y(x, y, z)\Delta y]}{\Delta x \Delta y} \\
&= \lim_{\Delta x \to 0} \frac{[E_y(x + \Delta x, y, z) - E_y(x, y, z)]}{\Delta x} \\
&= \partial_x E_y. \tag{1.31}
\end{aligned}
$$

The procedure for sides III and IV is the same, with the interchange of x and y, and a change of overall sign, leading to

$$(\mathbf{curl}\,\mathbf{E})_z = \partial_x E_y - \partial_y E_x \tag{1.32}$$

for the z component of $\mathbf{curl}\,\mathbf{E}$. The other components of $\mathbf{curl}$ follow in the same way with the cyclic substitution $x \to y$, $y \to z$, $z \to x$, as illustrated in Fig. 1.7.

With these substitutions, Eq. (1.32) can be extended to

$$(\mathbf{curl}\,\mathbf{E})_z = \partial_x E_y - \partial_y E_x, \quad \text{and cyclic.} \tag{1.33}$$

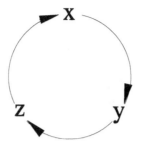

Figure 1.7: Cyclic order of $x \to y \to z$.

The "and cyclic" in Eq. (1.33) means that it stands for three equations, the one written and two others that follow by cyclic substitution. As a mnemonic, the positive term in any component of **curl** is written in cyclic order, and then the second term is in the opposite order with a minus sign.

The form of Eq. (1.33) suggests that **curl E** can be written as a cross product of the derivative operator ∇ and **E**:

$$\mathbf{curl\,E} = \nabla \times \mathbf{E}. \qquad (1.34)$$

Equation (1.34) is stated in words as "Curl E equals del cross E."

1.3.1 Stokes' Theorem

Equation (1.25) defines the average of **curl** over a finite surface. That equation can be rewritten as

$$\int_S \mathbf{dA} \cdot (\nabla \times \mathbf{F}) = \oint_C \mathbf{dr} \cdot \mathbf{F} \qquad (1.35)$$

for a vector **F**. In this form, it is called **Stokes' theorem** relating the integral of the curl of a vector over a surface S to the line integral of the vector around the closed curve C bounding the surface. From Stokes' theorem it follows that, if $\nabla \times \mathbf{F} = 0$ everywhere in a region, then

$$\oint_C \mathbf{dr} \cdot \mathbf{F} = 0 \qquad (1.36)$$

around any closed path in the region.

We now have seen three general properties of the electric field vector **E**. It is

- **irrotational,**

- **conservative,**

- **derivable from a potential.**

Although we have used the electric field as a simple example, these three properties also hold for a large class of vector fields. Some examples are the velocity vector of streamline fluid flow, the heat flow vector (with the temperature being the corresponding scalar field), the magnetic field **B** in a region with no current, and the gravitational field. The three properties have different physical manifestations, but they are mathematically equivalent. The connection between the properties of **E** are illustrated below:

IRROTATIONAL   **CONSERVATIVE**   **DERIVABLE FROM A POTENTIAL**

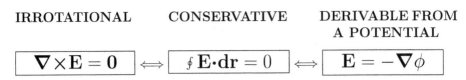

$$\boxed{\boldsymbol{\nabla}\times\mathbf{E}=\mathbf{0}} \iff \boxed{\oint \mathbf{E}\cdot\mathbf{dr}=0} \iff \boxed{\mathbf{E}=-\boldsymbol{\nabla}\phi}$$

The arrows indicate that each property is directly derivable from the adjacent property.

1.4   Summary of Operations by $\boldsymbol{\nabla}$

We list here a summary of the application of the vector differential operator $\boldsymbol{\nabla}$. The results listed have been derived earlier in the text, or will be answers to problems. In most applications, it is useful to know the operation of $\boldsymbol{\nabla}$ on simple functions of the position vector **r**. These results are summarized below:

1. $\boldsymbol{\nabla} f(r) = \hat{\mathbf{r}}\frac{df}{dr}$

2. $\boldsymbol{\nabla}(\mathbf{p}\cdot\mathbf{r})|_{\mathbf{p}\,\text{constant}} = \mathbf{p}$

3. $(\mathbf{p}\cdot\boldsymbol{\nabla})\mathbf{r} = \mathbf{p}$

4. $\boldsymbol{\nabla}\cdot\mathbf{r} = 3$

5. $\boldsymbol{\nabla}\times\mathbf{r} = 0$

6. $\boldsymbol{\nabla}\cdot\left(\frac{\mathbf{r}}{r^3}\right) = 4\pi\delta(\mathbf{r})$ (will be derived in chapter 4)

1.5   Problems

1. Show that $\nabla(\mathbf{r}\cdot\mathbf{p}) = \mathbf{p}$, for a constant vector $\mathbf{p}$.

2. The potential energy between two electric dipoles, a distance $r > 0$ apart is

$$U = \frac{[3(\mathbf{p}\cdot\hat{\mathbf{r}})(\mathbf{p}'\cdot\hat{\mathbf{r}}) - \mathbf{p}\cdot\mathbf{p}']}{r^3}.$$

Find the force between the dipoles, given by $\mathbf{F} = -\nabla U$.

3. Find the divergence of $\mathbf{E} = q\mathbf{r}/r^3$ for $r > 0$.

4. Show explicitly that the curl of $\mathbf{E} = q\mathbf{r}/r^3$ vanishes.

5. The potential of an electric dipole $\mathbf{p}$ is given by

$$\phi(\mathbf{r}) = \frac{\mathbf{p}\cdot\mathbf{r}}{r^3}.$$

Find the electric field of the dipole (for $r > 0$), given by $\mathbf{E} = -\nabla\phi$.

Chapter 2

Vector Identities

2.1 Algebraic Identities

Two useful algebraic identities (to be given as problems) that we will use in expanding vector derivatives are:

1. The triple scalar product of three vectors $\mathbf{a}\cdot(\mathbf{b}\times\mathbf{c})$ has the symmetry properties

$$\mathbf{a}\cdot(\mathbf{b}\times\mathbf{c}) = \mathbf{b}\cdot(\mathbf{c}\times\mathbf{a}) = \mathbf{c}\cdot(\mathbf{a}\times\mathbf{b}) = (\mathbf{a}\times\mathbf{b})\cdot\mathbf{c}. \qquad (2.1)$$

   That is, it is invariant under cyclic permutation (in either direction) or the interchange of the dot and the cross.

2. The triple vector product $\mathbf{a}\times(\mathbf{b}\times\mathbf{c})$ can be expanded as

$$\mathbf{a}\times(\mathbf{b}\times\mathbf{c}) = \mathbf{b}(\mathbf{a}\cdot\mathbf{c}) - \mathbf{c}(\mathbf{a}\cdot\mathbf{b}). \qquad (2.2)$$

   This identity should be kept firmly in memory as the **bac minus cab** rule, using the mnemonic "a cross b cross c equals bac minus cab."

2.2 Properties of ∇

Operations on combinations of functions of position can be simplified by using the two distinct properties of ∇:

1. ∇ **is a differential operator.**

2. ∇ **is a vector.**

Because ∇ is a differential operator, it acts on functions one at a time, just as in d(uv)=udv+vdu. We also follow the convention that the differential operator

15

acts only on functions to its right, so the order in which ∇ appears must be to the left of the functions it acts on and to the right of the other functions. As a vector, ∇ must behave in any expansion like any other vector.

In every case, strict adherence to these two properties of ∇ will lead to the correct evaluation of vector derivatives. We give several examples below of the use of the two properties of ∇ and the algebraic identities.

2.3   Use of bac $-$ cab

The bac minus cab rule is used in expanding $\nabla\times(\mathbf{A}\times\mathbf{B})$, the curl of a cross product. We go through the derivation step by step to illustrate the method. Since ∇ is a differential operator, there will be two terms, one with $\mathbf{A}$ constant and one with $\mathbf{B}$ constant:

$$\nabla\times(\mathbf{A}\times\mathbf{B}) = \nabla\times(\mathbf{A}\times\mathbf{B})|_{\mathbf{A}\,\text{constant}} + \nabla\times(\mathbf{A}\times\mathbf{B})|_{\mathbf{B}\,\text{constant}}. \quad (2.3)$$

We treat these two terms separately. For $\mathbf{A}$ constant, the first step is to write the three vectors in the order $\mathbf{A}\,\nabla\,\mathbf{B}$, since ∇ only acts on $\mathbf{B}$. This is written down twice, anticipating the use of the bac minus cab rule, giving

$$\nabla\times(\mathbf{A}\times\mathbf{B}) = \mathbf{A}\,\nabla\,\mathbf{B} \quad \mathbf{A}\,\nabla\,\mathbf{B}. \quad (2.4)$$

The second step is to write the bac minus cab equation above the line, so this intermediate step looks like:

$$\begin{aligned} \mathbf{a}\times(\,\mathbf{b}\times\mathbf{c}) &= \mathbf{b}(\mathbf{a}\cdot\mathbf{c}) - \mathbf{c}(\mathbf{a}\cdot\mathbf{b}) \\ \nabla\times(\mathbf{A}\times\mathbf{B})|_{\mathbf{A}\,\text{constant}} &= \mathbf{A}\,\nabla\,\mathbf{B} \quad \mathbf{A}\,\nabla\,\mathbf{B}. \end{aligned} \quad (2.5)$$

(Really write down the 'bac-cab' as shown here to avoid making an error in this step.)

We use the left hand side of this step to relate the a, b, and c of the bac minus cab rule to the vectors ∇, $\mathbf{A}$, and $\mathbf{B}$, so $\mathbf{a}\sim\nabla$, $\mathbf{b}\sim\mathbf{A}$, $\mathbf{c}\sim\mathbf{B}$. This association tells us where to put the dots and crosses on the right hand side, in order to preserve the vector algebra. This third step leaves

$$\begin{aligned} \mathbf{a}\times(\mathbf{b}\times\mathbf{c}) &= \mathbf{b}\,(\mathbf{a}\cdot\mathbf{c}) - \mathbf{c}(\mathbf{a}\cdot\mathbf{b}) \\ \nabla\times(\mathbf{A}\times\mathbf{B})|_{\mathbf{A}\,\text{constant}} &= \mathbf{A}(\nabla\cdot\mathbf{B}) - (\mathbf{A}\cdot\nabla)\mathbf{B}, \end{aligned} \quad (2.6)$$

with the sign determined by bac minus cab.

There will be two more terms for $\mathbf{B}$ constant. These could be found by repeating the same process as when $\mathbf{A}$ was held constant, but there is a simpler way. We note that the original expression $\nabla\times(\mathbf{A}\times\mathbf{B})$ changes sign under the interchange $\mathbf{A} \rightleftharpoons \mathbf{B}$. Then, the two terms with $\mathbf{B}$ constant just follow by interchanging $\mathbf{A}$ and $\mathbf{B}$ with a sign change in Eq. (2.6). This gives the four terms in the full expansion:

$$\nabla \times (\mathbf{A} \times \mathbf{B}) = \mathbf{A}(\nabla \cdot \mathbf{B}) - (\mathbf{A} \cdot \nabla)\mathbf{B} - \mathbf{B}(\nabla \cdot \mathbf{A}) + (\mathbf{B} \cdot \nabla)\mathbf{A}. \qquad (2.7)$$

The expression $(\mathbf{A} \cdot \nabla)$ in the above equation represents the magnitude $|\mathbf{A}|$ times the directional derivative $\hat{\mathbf{A}} \cdot \nabla$, which was defined just after Eq. (1.2). In this case, the directional derivative is acting on a vector rather than a scalar.

A useful identity involving the directional derivative is $(\mathbf{A} \cdot \nabla)\mathbf{r} = \mathbf{A}$. This can be derived by considering $(\mathbf{A} \cdot \nabla)\mathbf{r}$ as the bac of bac-cab. First, we write

$$(\mathbf{A} \cdot \nabla)\mathbf{r} \;=\; \mathbf{A} \; \nabla \; \mathbf{r} \quad \mathbf{A} \; \nabla \; \mathbf{r},$$

and then

$$\mathbf{b}(\mathbf{a} \cdot \mathbf{c}) \;=\; \mathbf{a} \times (\mathbf{b} \times \mathbf{c}) + \mathbf{c}(\mathbf{a} \cdot \mathbf{b})$$
$$(\mathbf{A} \cdot \nabla)\mathbf{r} \;=\; \mathbf{A} \; \nabla \; \mathbf{B} \quad \mathbf{A} \nabla \; \mathbf{B}.$$

This shows us that $\mathbf{a} \sim \mathbf{A}$, $\mathbf{b} \sim \mathbf{r}$, $\mathbf{c} \sim \nabla$, so

$$(\mathbf{A} \cdot \nabla)\mathbf{r} \;=\; -\mathbf{A} \times (\nabla \times \mathbf{r}) + \nabla(\mathbf{A} \cdot \mathbf{r})|_{\mathbf{A}\,\text{constant}} = \mathbf{A}. \qquad (2.8)$$

Strict application of the bac minus cab rule would have given the first term on the right hand side of Eq. (2.8) as $\mathbf{A} \times (\mathbf{r} \times \nabla)$. Since the ∇ must be to the left of the $\mathbf{r}$, we have interchanged their order in the cross product, giving the minus sign in that term. Then we used the fact that $\nabla \times \mathbf{r} = \mathbf{0}$. Also note that, although $\mathbf{A}$ need not be a constant vector for the identity to hold, we took it as constant in the term $\nabla(\mathbf{A} \cdot \mathbf{r})$ because the ∇ should not be acting on the $\mathbf{A}$ in that term.

We can also use bac minus cab to derive the identity

$$\nabla(\mathbf{A} \cdot \mathbf{B}) = \mathbf{A} \times (\nabla \times \mathbf{B}) + (\mathbf{A} \cdot \nabla)\mathbf{B} + \mathbf{B} \times (\nabla \times \mathbf{A}) + (\mathbf{B} \cdot \nabla)\mathbf{A}. \qquad (2.9)$$

We consider $\nabla(\mathbf{A} \cdot \mathbf{B})$ to be the bac of bac minus cab, so the intermediate step for $\mathbf{A}$ constant is

$$\mathbf{b}(\mathbf{a} \cdot \mathbf{c}) \qquad\quad = \;\; \mathbf{a} \times (\mathbf{b} \times \mathbf{c}) + \mathbf{c}(\mathbf{a} \cdot \mathbf{b})$$
$$\nabla(\mathbf{A} \cdot \mathbf{B})|_{\mathbf{A}\,\text{constant}} = \;\; \mathbf{A} \; \nabla \; \mathbf{B} \quad \mathbf{A} \; \nabla \; \mathbf{B}. \qquad (2.10)$$

The left hand side of this equation shows that the appropriate identification is $\mathbf{a} \sim \mathbf{A}$, $\mathbf{b} \sim \nabla$, $\mathbf{c} \sim \mathbf{B}$, which tells us how to put in the dots and crosses on the right hand side, so

$$\nabla(\mathbf{A} \cdot \mathbf{B})|_{\mathbf{A}\,\text{constant}} = \mathbf{A} \times (\nabla \times \mathbf{B}) + (\mathbf{A} \cdot \nabla)\mathbf{B}. \qquad (2.11)$$

The other two terms with $\mathbf{B}$ constant follow by just interchanging $\mathbf{A}$ and $\mathbf{B}$ since the original expression $\boldsymbol{\nabla}(\mathbf{A\cdot B})$ is symmetric in $\mathbf{A}$ and $\mathbf{B}$.

It is important to implement the above steps carefully in the order shown in the examples, leaving nothing out. This is especially vital while you are getting used to the procedure. Shortcuts will lead to error.

2.4 Problems

1. Derive the algebraic vector identities

    (a) $\mathbf{a}\cdot(\mathbf{b}\times\mathbf{c}) = \mathbf{b}\cdot(\mathbf{c}\times\mathbf{a}) = \mathbf{c}\cdot(\mathbf{a}\times\mathbf{b}) = (\mathbf{a}\times\mathbf{b})\cdot\mathbf{c}$.

    (b) $\mathbf{a}\times(\mathbf{b}\times\mathbf{c}) = \mathbf{b}(\mathbf{a}\cdot\mathbf{c}) - \mathbf{c}(\mathbf{a}\cdot\mathbf{b})$.

2. Calculate $\nabla\times(\mathbf{B}\times\mathbf{r})$ with $\mathbf{B}$ a constant vector.

3. The vector potential of a magnetic dipole $\boldsymbol{\mu}$ is given by

$$\mathbf{A}(\mathbf{r}) = \frac{\boldsymbol{\mu}\times\mathbf{r}}{r^3}.$$

    (a) Calculate the divergence of $\mathbf{A}$.

    (b) Find the magnetic field of the dipole (for $r > 0$), given by $\mathbf{B} = \nabla\times\mathbf{A}$.

4. Calculate the curl and divergence of $\mathbf{B}$, found in the previous problem.

5. Find the curl and divergence of each of the vector fields:

    (a) $\mathbf{F} = (\mathbf{r}\times\mathbf{p})(\mathbf{r}\cdot\mathbf{p})$,

    (b) $\mathbf{G} = (\mathbf{r}\cdot\mathbf{p})^2\mathbf{r}$.

    The vector $\mathbf{p}$ is a constant vector.

6. Show that the operator $\mathbf{L} = -i\mathbf{r}\times\nabla$ satisfies $\mathbf{L}\times\mathbf{L}\phi = i\mathbf{L}\phi$. [Hint: Let $\mathbf{E} = -\nabla\phi$. Then expand $(\mathbf{r}\times\nabla)\times(\mathbf{r}\times\mathbf{E})$ in bac-cab.]

Chapter 3

Integral Theorems

3.1 Green's Theorems

We can use the divergence theorem to derive two theorems, first derived by George Green in 1828. Green's theorems are of great use in solving vector differential equations like Poisson's equation and Laplace's equation.

Applying the divergence theorem to the combination $\phi\boldsymbol{\nabla}\psi$ of two scalar fields gives

$$\oint_S \mathbf{dA}\cdot[\phi\boldsymbol{\nabla}\psi] = \int_V \boldsymbol{\nabla}\cdot[\phi\boldsymbol{\nabla}\psi]d^3r. \tag{3.1}$$

Then, expanding $\boldsymbol{\nabla}\cdot[\phi\boldsymbol{\nabla}\psi]$ using the derivative property of $\boldsymbol{\nabla}$ gives **Green's first theorem**:

$$\oint_S \mathbf{dA}\cdot[\phi\boldsymbol{\nabla}\psi] = \int_V [\phi\nabla^2\psi + (\boldsymbol{\nabla}\phi)\cdot(\boldsymbol{\nabla}\psi)]d^3r. \tag{3.2}$$

Interchanging ϕ and ψ in Eq. (3.2), and subtracting leads to **Green's second theorem**:

$$\oint_S \mathbf{dA}\cdot[\phi\boldsymbol{\nabla}\psi - \psi\boldsymbol{\nabla}\phi] = \int_V [\phi\nabla^2\psi - \psi\nabla^2\phi]d^3r. \tag{3.3}$$

Green's second theorem could also be derived by applying the divergence theorem to the combination $[\phi\boldsymbol{\nabla}\psi - \psi\boldsymbol{\nabla}\phi]$. Green's second theorem is used more often than his first theorem. If someone just refers to "Green's theorem," the reference is usually to Green's second theorem.

Green's theorems involve the **Laplacian** differential operator ∇^2, defined by the application of the divergence to the gradient of a scalar. In Cartesian coordinates, the Laplacian operator is given by

$$\nabla^2 = \partial_x^2 + \partial_y^2 + \partial_z^2, \tag{3.4}$$

21

but it is more complicated in other coordinate systems.

Although these derivations of Green's two theorems seem almost trivial, they were derived by Green long before the development of the divergence theorem or the use of compact vector notation. Green worked them out in isolation, and his developments in vector calculus were not discovered until 1845 by William Thomson (later to be Lord Kelvin). At that time, they provided a monumental advance in the development of physics.

3.2   Gradient Theorem

Also using the divergence theorem, we can derive a **gradient theorem** relating the integral over a volume V of the gradient of a scalar to the integral over the bounding surface S of the scalar function:

$$\int_V d^3r \boldsymbol{\nabla}\phi = \oint_S \mathbf{dA}\phi. \tag{3.5}$$

The proof of this theorem follows by first dotting the left hand side of Eq. (3.5) by an arbitrary constant vector $\mathbf{k}$

$$
\begin{aligned}
\mathbf{k} \cdot \int_V d^3r \boldsymbol{\nabla}\phi &= \int_V d^3r \boldsymbol{\nabla}\cdot(\mathbf{k}\phi), &&\text{(since $\mathbf{k}$ is constant)} \\
&= \oint_S \mathbf{dA}\cdot\mathbf{k}\phi, &&\text{(by the divergence theorem)} \\
&= \mathbf{k} \cdot \oint_S \mathbf{dA}\phi &&\text{(since $\mathbf{k}$ is constant).}
\end{aligned}
\tag{3.6}
$$

Now, since $\mathbf{k}$ is an arbitrary vector, Eq. (3.6) must hold for any component of the vectors (in any coordinate system), and the vector equation (3.5) follows.

3.3   Curl Theorem

A **curl theorem**:

$$\int_V d^3r \boldsymbol{\nabla}\times\mathbf{E} = \oint_S \mathbf{dA}\times\mathbf{E} \tag{3.7}$$

can be proved in the same way as was the gradient theorem by dotting with an arbitrary constant vector $\mathbf{k}$:

$$
\begin{aligned}
\mathbf{k} \cdot \int_V d^3r \boldsymbol{\nabla}\times\mathbf{E} &= \int_V d^3r \mathbf{k}\cdot(\boldsymbol{\nabla}\times\mathbf{E}) \\
&= \int_V d^3r \boldsymbol{\nabla}\cdot(\mathbf{E}\times\mathbf{k}) \\
&= \oint_S \mathbf{dA}\cdot(\mathbf{E}\times\mathbf{k}) \\
&= \mathbf{k} \cdot \oint_S \mathbf{dA}\times\mathbf{E}.
\end{aligned}
\tag{3.8}
$$

Since **k** is an arbitrary vector, it follows, as before, that the vector equation (3.7) must follow from Eq. (3.8). In the steps in Eq. (3.8), we moved **k** in and out of integrals and derivatives because it is a constant vector. Notice that, at each step, the vector character of the equation is preserved. In two of the steps, we used the cyclic property of the triple scalar product.

3.4   Integral Definition of Del

Taking the limit of the divergence, gradient, or curl theorem as the volume shrinks to a point, we can provide a definition of ∇ that is independent of any coordinate system:

$$\nabla = \lim_{V \to 0} \frac{1}{V} \oint_S d\mathbf{A}. \tag{3.9}$$

In Eq. (3.9), the vector derivative operator ∇ is defined in terms of the vector integral operator on the right hand side of the equation. Then, any operation of the vector differential operator ∇ can be calculated by the corresponding operation by the vector integral operator.

Although it may seem unusual to define a derivative by an integral, it is really nothing new. The definition of the one-dimensional derivative is

$$\frac{df}{dx} = \lim_{\Delta x \to 0} \frac{f(x + \Delta x) - f(x)}{\Delta x}. \tag{3.10}$$

The numerator of this equation can be considered as the surface terms of a one-dimensional volume, Δx. In this sense, Eq. (3.10) can be considered the one-dimensional analog of the integral definition of ∇.

3.5   Stokes' Theorem for the Gradient

From Stokes' theorem, we can derive a **Stokes' theorem for the gradient**:

$$\int_S d\mathbf{A} \times (\nabla \phi) = \oint_C d\mathbf{r}\phi. \tag{3.11}$$

The proof of this theorem is similar to that for the gradient and curl theorems. We dot the left hand side of Eq. (3.11) by an arbitrary constant vector **k**:

$$\begin{aligned}
\mathbf{k} \cdot \int_S d\mathbf{A} \times (\nabla \phi) &= \int_S \mathbf{k} \cdot [d\mathbf{A} \times (\nabla \phi)] \\
&= \int_S d\mathbf{A} \cdot [(\nabla \phi) \times \mathbf{k}] \\
&= \int_S d\mathbf{A} \cdot [\nabla \times (\mathbf{k}\phi)]
\end{aligned}$$

$$= \oint_C d\mathbf{r} \cdot (\mathbf{k}\phi)$$

$$= \mathbf{k} \cdot \oint_C d\mathbf{r}\phi, \tag{3.12}$$

and again the vector equation follows because $\mathbf{k}$ is an arbitrary vector.

3.6   Gauss's Law

A method for using symmetry to find the electric field $\mathbf{E}$ without integrating over the charge distribution is given by **Gauss's law**. We can derive Gauss's law from Coulomb's law for a single point charge. We start with the surface integral

$$\oint_S d\mathbf{A} \cdot \mathbf{E} = q \oint_S d\mathbf{A} \cdot \left(\frac{\mathbf{r}}{r^3}\right) \tag{3.13}$$

of the normal component of $\mathbf{E}$ over a closed surface surrounding the point charge, as shown in Fig. 3.1.

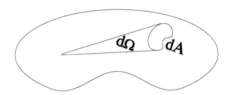

Figure 3.1: The solid angle $d\Omega$ subtended by the surface differential dA for Gauss's law.

The vector differential of area $d\mathbf{A}$ is an infinitesimal surface element of magnitude dA. Since it is infinitesimal, it approaches a plane surface, tangent to the closed surface. By convention, its vector direction is along the outward normal to the closed surface. The integrand of the surface integral in Eq. (3.13) can be recognized as the definition of the solid angle subtended by the differential surface element $d\mathbf{A}$ as can be seen on Fig. 3.1:

$$d\Omega = \frac{\hat{\mathbf{r}} \cdot d\mathbf{A}}{r^2}. \tag{3.14}$$

Then the surface integral can be written as

$$\oint_S d\mathbf{A} \cdot \mathbf{E} = q \oint d\Omega = 4\pi\, q, \tag{3.15}$$

with the factor 4π arising as the magnitude of the total solid angle of any closed surface.

If the point charge q were located outside the closed surface, the surface integral would be zero. This follows from applying the divergence theorem to the surface integral in Eq. (3.13):

$$\oint_S d\mathbf{A} \cdot \left(\frac{\mathbf{r}}{r^3}\right) = \int_V \left[\boldsymbol{\nabla} \cdot \left(\frac{\mathbf{r}}{r^3}\right)\right] d^3r = 0 \qquad (3.16)$$

because $\boldsymbol{\nabla}\cdot(\mathbf{r}/r^3) = 0$ in the volume. (See Problem 1.3.) Then, the integral over the closed surface will be $4\pi q$ for a point charge inside the surface, and zero for a point charge outside the surface.

For a collection of point charges, only those inside the surface will contribute to the integral, and we have Gauss's law

$$\oint_S d\mathbf{A}\cdot\mathbf{E} = 4\pi\, Q_{\text{enclosed}}, \qquad (3.17)$$

where Q_{enclosed} is the net charge within the surface. Since a continuous charge distribution is really a very large number of point charges, Gauss's law for a continuous distribution is also given by Eq. (3.17), with

$$Q_{\text{enclosed}} = \int \rho(\mathbf{r})d^3r, \qquad (3.18)$$

where the integral is over the volume enclosed by the closed surface.

Gauss's law provides a powerful and simple method to find $\mathbf{E}$ whenever there is enough symmetry to enable the surface integral to be done without integration. But, if use of Gauss's law requires a complicated surface integration, then another method should be used to find $\mathbf{E}$.

Although we have derived Gauss's law for the electric field, it can also be applied to the gravitational field and other physical cases where symmetry can be used to simplify the surface integral. Some applications of Gauss's law are given as problems.

As a simple example of Gauss's law, we use it to derive Coulomb's law for a point charge, demonstrating the steps used in applying Gauss's law. The first step is to recognize the symmetry of the charge configuration, which, in the case of an isolated point charge is spherical symmetry about the point charge. The type of symmetry dictates the **Gaussian surface** to be used for the surface integral.

For the point charge, this is a sphere, of any radius r, centered at the point charge so as to make use of the symmetry, as shown in Fig. 3.3. Note that the Gaussian surface is just a mathematical surface that need not be (and usually isn't) any physical surface of the problem.

Next the symmetry is used to make simplifying observations about the $\mathbf{E}$ field at the Gaussian surface. For the point charge, we first observe that $\mathbf{E}$ must

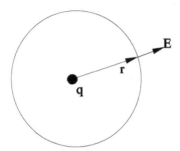

Figure 3.2: Gaussian sphere for a single point charge.

be in the radial direction with respect to the charge. This follows, most simply from the principle of insufficient reason. That is, looking at Fig. 3.3, there is no more reason for **E** to be directed to the right of radial than to the left, since there is no reference point other than the charge to define left or right.

Another, more mathematical, derivation of the radial direction of **E** is the fact that only one vector, the position vector **r**, can be defined for this geometry. Thus the vector function **E** must be given only in terms of this vector and can only be $\mathbf{E}(\mathbf{r}) = E(\mathbf{r})\hat{\mathbf{r}}$. (This reasoning will be used later in the book to simplify other vector integrations.) Figure 3.3 also shows that there is no preferred direction so **E** cannot depend on direction, and can be wriiten as $E(r)\hat{\mathbf{r}}$.

Now we can write the Gauss's law integral

$$\oint_S d\mathbf{A}\cdot\mathbf{E} = \oint E(r)r^2 d\Omega = 4\pi r^2 E(r) = 4\pi q. \tag{3.19}$$

This gives

$$\mathbf{E}(\mathbf{r}) = \frac{q\hat{\mathbf{r}}}{r^2} \tag{3.20}$$

for the electric field of a point charge.

3.7 Problems

1. Demonstrate the divergence theorem

$$\int_V d^3r \, \nabla \cdot \mathbf{F} \;=\; \oint_S d\mathbf{A} \cdot \mathbf{F}$$

using the function $\mathbf{F} = \mathbf{r}(\hat{\mathbf{p}} \cdot \mathbf{r})^2$ in a sphere of radius R.

2. Demonstrate the gradient theorem

$$\int_V d^3r \, \nabla \phi \;=\; \oint_S d\mathbf{A} \phi$$

using the function $\phi = r(\mathbf{r} \cdot \hat{\mathbf{p}})$ in a sphere of radius R.

3. Demonstrate the curl theorem

$$\int_V d^3r \, \nabla \times \mathbf{F} \;=\; \oint_S d\mathbf{A} \times \mathbf{F}$$

using the function $\mathbf{F} = \hat{\mathbf{q}} \times (\hat{\mathbf{p}} \times \mathbf{r})$ in a sphere of radius R.

4. (a) Use Gauss's law to find the electric field inside and outside a uniformly charged hollow sphere of charge Q and radius R.

   (b) Integrate $\mathbf{E}$ to find the potential inside and outside the hollow sphere.

5. (a) Use Gauss's law to find the electric field inside and outside a uniformly charged solid sphere of charge Q and radius R.

   (b) Integrate $\mathbf{E}$ to find the potential inside and outside the solid sphere.

6. (a) Use Gauss's law to find the electric field inside and outside a long straight wire of radius R with uniform charge density ρ.

   (b) Integrate $\mathbf{E}$ to find the potential inside and outside the wire with the boundary condition $\phi(R) = 0$.

7. Apply Gauss's law to an infinitesimal volume, and use the definition of divergence to derive the relation

$$\nabla \cdot \mathbf{E} \;=\; 4\pi\rho.$$

Chapter 4

Dirac Delta Function

4.1 Definition of Dirac Delta Function

To introduce the need for the Dirac delta function, we study the behavior at the origin of the divergence of the electric field of a point charge (or of the gravitational field of a point mass, or of any vector field that goes like $1/r^2$ as r approaches zero).

For $r > 0$, the divergence of the electric field of a point charge is given by

$$
\begin{aligned}
\boldsymbol{\nabla} \cdot \left(\frac{q\mathbf{r}}{r^3} \right) &= q \frac{\boldsymbol{\nabla} \cdot \mathbf{r}}{r^3} + q \mathbf{r} \cdot \boldsymbol{\nabla} \left(\frac{1}{r^3} \right) \\
&= \frac{3q}{r^3} - \frac{3q(\mathbf{r} \cdot \hat{\mathbf{r}})}{r^4} = 0, \quad r > 0.
\end{aligned}
\tag{4.1}
$$

The restriction $r > 0$ is necessary because both terms in Eq. (4.1) are singular at $r = 0$.

We now investigate the behavior of $\boldsymbol{\nabla} \cdot (\mathbf{r}/r^3)$ at $r = 0$. In fact, something quite dramatic happens at the origin to the divergence of $\mathbf{r}/r^3$. We can see this by applying the divergence theorem to $\mathbf{r}/r^3$:

$$
\int d^3 r \, \boldsymbol{\nabla} \cdot \left(\frac{\mathbf{r}}{r^3} \right) = \oint_S \frac{d\mathbf{A} \cdot \mathbf{r}}{r^3} = \oint d\Omega = 4\pi,
\tag{4.2}
$$

for integration over any volume containing the origin. Thus, even though $\boldsymbol{\nabla} \cdot (\mathbf{r}/r^3)$ vanishes at all but one point, its volume integral is not zero.

This property is consistent with one definition of the **Dirac delta function**, $\delta(\mathbf{r})$, in three dimensions:

$$
\begin{aligned}
\int_V d^3 r \, \delta(\mathbf{r}) &= 1, \quad \text{if } r = 0 \text{ inside V}, \\
\int_V d^3 r \, \delta(\mathbf{r}) &= 0, \quad \text{if } r \neq 0 \text{ inside V},
\end{aligned}
\tag{4.3}
$$

for any volume V. With this definition of $\delta(\mathbf{r})$, we see that

$$\nabla \cdot \left(\frac{\mathbf{r}}{r^3} \right) = 4\pi\delta(\mathbf{r}). \tag{4.4}$$

It follows from the vanishing of the integral of the delta function for any volume not containing the origin that

$$\delta(\mathbf{r}) = 0 \quad \text{if } r \neq 0. \tag{4.5}$$

To apply the divergence theorem to integrals over continuous functions, we extend Eq. (4.4) to

$$\nabla \cdot \left[\frac{(\mathbf{r} - \mathbf{r}')}{|\mathbf{r} - \mathbf{r}'|^3} \right] = 4\pi\delta(\mathbf{r} - \mathbf{r}'), \tag{4.6}$$

where the constant vector $\mathbf{r}'$ just shifts the origin of coordinates from $\mathbf{0}$ to $\mathbf{r}'$.

In any integral over a volume V containing the point $\mathbf{r}' = \mathbf{r}$, the region of integration can be shrunk to an infinitesimal volume surrounding $\mathbf{r}$. Then the integral including $\delta(\mathbf{r} - \mathbf{r}')$ has the **sifting property**

$$\begin{aligned} \int_V \delta(\mathbf{r} - \mathbf{r}')f(\mathbf{r}')d^3r' &= f(\mathbf{r}), \quad \text{if } \mathbf{r}' = \mathbf{r} \text{ inside V} \\ \int_V \delta(\mathbf{r} - \mathbf{r}')f(\mathbf{r}')d^3r' &= 0, \quad\;\; \text{if } \mathbf{r}' \neq \mathbf{r} \text{ inside V}, \end{aligned} \tag{4.7}$$

provided that $\lim_{\mathbf{r}' \to \mathbf{r}} f(\mathbf{r}')$ exists. Thus, an integral with a delta function is the simplest integral to do. The integration just involves evaluating the rest of the integrand at the point where the argument of the delta function vanishes.

Equation (4.7) is a somewhat better definition of the Dirac delta function than Eq. (4.3) because it permits a more general definition of the delta function with respect to a class of functions $f(\mathbf{r})$. Then, Eq. (4.3) follows from this definition if $f(\mathbf{r}')$ is chosen to be 1.

We must emphasize that the Dirac delta function is not a mathematical function in the strict sense. In fact, as a function, it does not make sense. It would vanish everywhere, except where it was not defined (loosely speaking, 'infinite'). That is why we have been careful, in either definition, to define the delta function only in terms of its property in integrals. When we write it in equations where we do not integrate, such as Eq. (4.6), it is always with the understanding that the delta function will only be given physical meaning in a subsequent integration. That is, in equations like Eq. (4.6), the delta function is just an indication of how to perform a pending integration.

4.2 Applications of the Dirac Delta Function

Taking the divergence of the Coulomb integral for the electric field of a volume distribution of charge leads to

$$
\begin{aligned}
\boldsymbol{\nabla}{\cdot}\mathbf{E}(\mathbf{r}) &= \boldsymbol{\nabla}{\cdot}\int \frac{(\mathbf{r}-\mathbf{r}')\rho(\mathbf{r}')d^3r'}{|\mathbf{r}-\mathbf{r}'|^3} \\
&= \int \rho(\mathbf{r}')d^3r'\boldsymbol{\nabla}{\cdot}\left[\frac{(\mathbf{r}-\mathbf{r}')}{|\mathbf{r}-\mathbf{r}'|^3|}\right] \\
&= 4\pi \int \rho(\mathbf{r}')d^3r'\delta(\mathbf{r}-\mathbf{r}').
\end{aligned}
\tag{4.8}
$$

Doing the delta function integral gives

$$
\boldsymbol{\nabla}{\cdot}\mathbf{E} = 4\pi\rho
\tag{4.9}
$$

for any continuous charge distribution.

Gauss's law can be derived by applying the divergence theorem to Eq. (4.9)

$$
\oint_S \mathbf{E}{\cdot}\mathbf{dA} = \int_V \boldsymbol{\nabla}{\cdot}\mathbf{E}d^3r = 4\pi \int_V \rho(\mathbf{r})d^3r = 4\pi Q_{\text{enclosed}}.
\tag{4.10}
$$

Gauss's law holds for any vector field, including the gravitational field as well as the electric field, that is given by the first integral in Eq. (4.8).

Equation (4.9) can be put in terms of the potential ϕ, leading to **Poisson's equation**[1]

$$
-\boldsymbol{\nabla}{\cdot}\mathbf{E} = \boldsymbol{\nabla}{\cdot}(\boldsymbol{\nabla}\phi) = \nabla^2\phi = -4\pi\rho.
\tag{4.11}
$$

The homogeneous form of Poisson's equation, with the source function $\rho = 0$,

$$
\nabla^2\phi = 0,
\tag{4.12}
$$

is called **Laplace's equation**.

4.3 Singularities of Dipole Fields

Another occurrence of the Dirac delta function is in the singular behavior at the origin of the electromagnetic field of a point electric or magnetic dipole. The potential of a point electric dipole is given by

$$
\phi_{\mathbf{p}}(\mathbf{r}) = \frac{\mathbf{r}{\cdot}\mathbf{p}}{r^3}.
\tag{4.13}
$$

---

[1]Poisson's equation is often defined in mathematics books without the factor 4π. Then a factor of $(1/4\pi)$ would appear in the solution of the equation.

The electric field of a point electric dipole is given by the negative gradient of the potential:

$$
\begin{aligned}
\mathbf{E_p}(\mathbf{r}) &= -\boldsymbol{\nabla}\phi_{\mathbf{p}}(\mathbf{r}) = -\boldsymbol{\nabla}\left(\frac{\mathbf{p}\cdot\mathbf{r}}{r^3}\right) \\
&= -\frac{\mathbf{p}}{r^3} - (\mathbf{p}\cdot\mathbf{r})\boldsymbol{\nabla}\left(\frac{1}{r^3}\right).
\end{aligned}
\tag{4.14}
$$

The term $\boldsymbol{\nabla}(1/r^3)$ has to be treated carefully because of its singular nature. We do this by relating it to the known result

$$
\boldsymbol{\nabla}\cdot\left(\frac{\mathbf{r}}{r^3}\right) = 4\pi\delta(\mathbf{r}).
\tag{4.15}
$$

Since the gradient in $\boldsymbol{\nabla}(1/r^3)$ acts on a function of r, it can be written as

$$
\begin{aligned}
\boldsymbol{\nabla}\left(\frac{1}{r^3}\right) &= \hat{\mathbf{r}}\left[\hat{\mathbf{r}}\cdot\boldsymbol{\nabla}\left(\frac{1}{r^3}\right)\right] = \frac{\hat{\mathbf{r}}}{r}\left[(\mathbf{r}\cdot\boldsymbol{\nabla})\left(\frac{1}{r^3}\right)\right] \\
&= \frac{\hat{\mathbf{r}}}{r}\left[\boldsymbol{\nabla}\cdot\left(\frac{\mathbf{r}}{r^3}\right) - \frac{\boldsymbol{\nabla}\cdot\mathbf{r}}{r^3}\right] \\
&= \frac{\hat{\mathbf{r}}}{r}\left[4\pi\delta(\mathbf{r}) - \frac{3}{r^3}\right].
\end{aligned}
\tag{4.16}
$$

Now we can continue Eq. (4.14) as

$$
\begin{aligned}
\mathbf{E_p}(\mathbf{r}) &= -\frac{\mathbf{p}}{r^3} - (\mathbf{p}\cdot\mathbf{r})\boldsymbol{\nabla}\left(\frac{1}{r^3}\right) \\
&= \frac{3(\mathbf{p}\cdot\hat{\mathbf{r}})\hat{\mathbf{r}} - \mathbf{p}}{r^3} - 4\pi\hat{\mathbf{r}}(\hat{\mathbf{r}}\cdot\mathbf{p})\delta(\mathbf{r}).
\end{aligned}
\tag{4.17}
$$

Equation (4.17) shows the dipole electric field with its singular behavior at the origin. The singular behavior at the origin adds a contact term to the potential energy of two dipoles, $\mathbf{p}$ and $\mathbf{p}'$. This is given by

$$
U_{\mathbf{pp}'} = -\mathbf{p}\cdot\mathbf{E_{p'}} = \frac{\mathbf{p}\cdot\mathbf{p}' - 3(\mathbf{p}\cdot\hat{\mathbf{r}})(\mathbf{p}'\cdot\hat{\mathbf{r}})}{r^3} + 4\pi(\mathbf{p}\cdot\hat{\mathbf{r}})(\hat{\mathbf{r}}\cdot\mathbf{p}')\delta(\mathbf{r}).
\tag{4.18}
$$

The contact term in the potential energy is not too important in classical physics where the likelihood of two dipoles touching is remote, but it must be taken into account in quantum mechanics whenever the wave function of the two dipoles does not vanish for $r = 0$. This only happens for s-wave (orbital angular momentum zero) states for which the wave function is spherically symmetric.

The potential energy is usually integrated with the wave function squared, so the delta function term enters averaged over all solid angles. The average over solid angle is defined as

$$
\langle(\hat{\mathbf{p}}\cdot\hat{\mathbf{r}})(\hat{\mathbf{r}}\cdot\hat{\mathbf{p}}')\rangle = \frac{1}{4\pi}\int d\Omega(\hat{\mathbf{p}}\cdot\hat{\mathbf{r}})(\hat{\mathbf{r}}\cdot\hat{\mathbf{p}}').
\tag{4.19}
$$

We evaluate the average over solid angle the following way. We first take the constant vector $\hat{\mathbf{p}}'$ out of the integral, and write

$$\langle(\hat{\mathbf{p}}\cdot\hat{\mathbf{r}})(\hat{\mathbf{r}}\cdot\hat{\mathbf{p}}')\rangle = \frac{1}{4\pi}\mathbf{I}\cdot\hat{\mathbf{p}}', \tag{4.20}$$

where $\mathbf{I}$ is a vector given by

$$\mathbf{I} = \int d\Omega(\hat{\mathbf{p}}\cdot\hat{\mathbf{r}})\hat{\mathbf{r}}. \tag{4.21}$$

Now we note a useful property of a vector integral like $\mathbf{I}$ that depends on only one constant vector. The integral can only be in the $\hat{\mathbf{p}}$ direction, since no other vector direction can be defined. Thus we can write $\mathbf{I}$ as

$$\mathbf{I} = I\hat{\mathbf{p}}. \tag{4.22}$$

Then,

$$\langle(\hat{\mathbf{p}}\cdot\hat{\mathbf{r}})(\hat{\mathbf{r}}\cdot\hat{\mathbf{p}}')\rangle = (\hat{\mathbf{p}}\cdot\hat{\mathbf{p}}')I, \tag{4.23}$$

where the scalar integral I is given by

$$I = \hat{\mathbf{p}}\cdot\mathbf{I} = \int d\Omega(\hat{\mathbf{p}}\cdot\hat{\mathbf{r}})(\hat{\mathbf{r}}\cdot\hat{\mathbf{p}}). \tag{4.24}$$

Choosing the $\hat{\mathbf{p}}$ direction as the z-axis, the angular integral becomes

$$I = \int d\Omega(z^2/r^2) = 4\pi\langle z^2/r^2\rangle. \tag{4.25}$$

There is spherical symmetry, so $\langle z^2\rangle = \langle x^2\rangle = \langle y^2\rangle$. Since $z^2 + x^2 + y^2 = r^2$, it follows that $\langle z^2/r^2\rangle = 1/3$, and we have the result

$$I = \int d\Omega(z^2/r^2) = 4\pi\langle z^2/r^2\rangle = 4\pi/3. \tag{4.26}$$

Thus the average over angles in Eq. (4.19) is

$$\langle(\hat{\mathbf{p}}\cdot\hat{\mathbf{r}})(\hat{\mathbf{r}}\cdot\hat{\mathbf{p}}')\rangle = \frac{1}{3}(\hat{\mathbf{p}}\cdot\hat{\mathbf{p}}'). \tag{4.27}$$

This result for the average over angles will be used throughout the book whenever the angular integration arises.

The potential energy, averaged over angles, is

$$\langle U_{\mathbf{p}\mathbf{p}'}\rangle = \frac{4\pi}{3}(\mathbf{p}\cdot\mathbf{p}')\delta(\mathbf{r} - \mathbf{r}'). \tag{4.28}$$

Note that the non-singular part of U vanishes in the average over solid angle, so Eq. (4.28) gives the full potential energy for a spherically symmetric quantum state of two electric dipoles.

The singular part of the electric field given by Eq. (4.17) can also be averaged over solid angle to give

$$\mathbf{E_p(r)} \quad = \quad \frac{3(\mathbf{p} \cdot \hat{\mathbf{r}})\hat{\mathbf{r}} - \mathbf{p}}{r^3} - \frac{4\pi}{3}\mathbf{p}\delta(\mathbf{r}). \tag{4.29}$$

This form for the singular part is more common than that of Eq. (4.17) because most other derivations of the singular part include the integral over solid angle as part of the derivation.

There is also a singular part of the magnetic field of a magnetic dipole. The vector potential of a magnetic dipole $\boldsymbol{\mu}$ is given by

$$\mathbf{A(r)} = \frac{\boldsymbol{\mu} \times \mathbf{r}}{r^3}, \tag{4.30}$$

and the magnetic field of the dipole is given by

$$\mathbf{B} = \boldsymbol{\nabla} \times \mathbf{A}. \tag{4.31}$$

The magnetic field is thus

$$\begin{aligned} \mathbf{B(r)} \quad &= \quad \boldsymbol{\nabla} \times \left(\frac{\boldsymbol{\mu} \times \mathbf{r}}{r^3}\right) \\ &= \quad \boldsymbol{\mu}\left[\boldsymbol{\nabla} \cdot \left(\frac{\mathbf{r}}{r^3}\right)\right] - (\boldsymbol{\mu} \cdot \boldsymbol{\nabla})\left(\frac{\mathbf{r}}{r^3}\right) \\ &= \quad 4\pi\boldsymbol{\mu}\delta(\mathbf{r}) - \frac{\boldsymbol{\mu}}{r^3} - \mathbf{r}\left[\boldsymbol{\mu} \cdot \boldsymbol{\nabla}\left(\frac{1}{r^3}\right)\right] \end{aligned} \tag{4.32}$$

We have seen in Eq. (4.16) that

$$\boldsymbol{\nabla}\left(\frac{1}{r^3}\right) \quad = \quad \frac{\hat{\mathbf{r}}}{r}\left[4\pi\delta(\mathbf{r}) - \frac{3}{r^3}\right]. \tag{4.33}$$

Thus, Eq. (4.32) leads to

$$\begin{aligned} \mathbf{B(r)} \quad &= \quad \frac{(3\boldsymbol{\mu}\cdot\hat{\mathbf{r}})\hat{\mathbf{r}} - \boldsymbol{\mu}}{r^3} + 4\pi\boldsymbol{\mu}\delta(\mathbf{r}) - 4\pi\frac{(\boldsymbol{\mu}\cdot\hat{\mathbf{r}})\hat{\mathbf{r}}}{r}\delta(\mathbf{r}) \tag{4.34} \\ &= \quad \frac{(3\boldsymbol{\mu}\cdot\hat{\mathbf{r}})\hat{\mathbf{r}} - \boldsymbol{\mu}}{r^3} + \frac{8\pi}{3}\boldsymbol{\mu}\delta(\mathbf{r}), \tag{4.35} \end{aligned}$$

where the singular term in the magnetic field has been averaged over solid angle in Eq, (4.35) The result is that a magnetic dipole has the same non-singular field distribution as does an electric dipole, but their singularities at the origin differ in sign and magnitude.

The delta function singularity in the magnetic field of a magnetic dipole leads to the hyperfine interaction in atoms due to the dipole-dipole force between atomic electrons and the nucleus of the atom. The hyperfine energy is given by

$$U = -\boldsymbol{\mu} \cdot \mathbf{B} = -\frac{8\pi}{3} \boldsymbol{\mu}_{\text{nucleus}} \cdot \boldsymbol{\mu}_{\text{electron}} \delta(\mathbf{r}). \tag{4.36}$$

The magnitude and sign of the measured hyperfine interaction agrees with the the singular term in Eq. (4.36), confirming that the magnetic field of the intrinsic magnetic moment of the electron is given by Eq. (4.35).

4.4 One-dimensional Delta Function

The delta function is defined in one dimension by the integral (with $a < b$)

$$\int_a^b \delta(x - x') f(x') dx' = f(x), \quad \text{if } a < x < b,$$
$$= 0, \quad \text{if } x < a, \text{ or } x > b, \tag{4.37}$$

provided that $\lim_{x' \to x} f(x')$ exists. This is called the **sifting property** of the one dimensional delta function. A somewhat simpler definition of the delta function is given if the function $f(x) = 1$. Then

$$\int_a^b \delta(x) dx = 1, \quad \text{if } a < 0 < b,$$
$$= 0, \quad \text{if } a > 0, \text{ or } b < 0. \tag{4.38}$$

The delta function can be represented by the first derivative of the unit step function, $\theta(x)$. The unit step function is defined by

$$\theta(x) = 0 \text{ if } x < 0$$
$$\theta(x) = 1 \text{ if } x > 0. \tag{4.39}$$

Since $\theta(x)$ is constant for $x \neq 0$, its derivative is zero for $x \neq 0$.

The integral of the derivative of the step function over the range (a, b) (if $a < 0 < b$) results in

$$\int_a^b \left[\frac{d\theta(x)}{dx} \right] f(x) dx = [\theta(x) f(x)]_a^b - \int_a^b \theta(x) \left[\frac{df(x)}{dx} \right] dx$$
$$= f(b) - \int_0^b \left[\frac{df(x)}{dx} \right] dx$$
$$= f(b) - f(b) + f(0) = f(0), \tag{4.40}$$

where we have integrated by parts. If x did not equal zero within the range (a, b), then the integration would result in zero.

The result of the integration shows that

$$\frac{d\theta(x)}{dx} = \delta(x),\tag{4.41}$$

so the first derivative of $\theta(x)$ provides a representation of the delta function. Other representations of the delta function are given in the problems.

4.5   Problems

1. For the potential $\phi(r) = qe^{-\mu r}/r$,

   (a) find the electric field.

   (b) find the charge distribution that produces this potential.

   (c) show that Gauss's law is satisfied by your answers. (Use a spherical Gaussian surface and be careful about the origin.)

2. Show that in the limit $a \to 0$, the function

$$f(r) \;=\; \frac{e^{-r^2/a^2}}{(\sqrt{\pi}a)^3}$$

represents the three-dimensional delta function, $\delta(\mathbf{r})$.

3. Show that in the limit $a \to 0$, the function

$$f(x) \;=\; \frac{e^{-x^2/a^2}}{a\sqrt{\pi}}$$

represents the one-dimensional delta function, $\delta(x)$.

4. Show that

$$f(x) \;=\; \lim_{\epsilon \to 0} \left[\mathrm{Im}\left(\frac{1}{x - i\epsilon} \right) \right] = \pi\delta(x).$$

5. Show that

$$\delta[f(x)] \;=\; \sum_i \frac{\delta(x - x_i)}{\left| \frac{df}{dx} \right|_{x=x_i}},$$

where the x_i are the zeros of $f(x)$.

Chapter 5

Green's Functions

5.1 Application of Green's Second Theorem

In this section we use Green's second theorem to find the solution to Poisson's equation with given boundary conditions. We start with Green's second theorem, written in terms of integrals over the variable $\mathbf{r}'$:

$$\int_V [\phi(\mathbf{r}')\boldsymbol{\nabla}'^2\psi(\mathbf{r}') - \psi(\mathbf{r}')\boldsymbol{\nabla}'^2\phi(\mathbf{r}')]d^3r' = \oint_S d\mathbf{A}'\cdot[\phi(\mathbf{r}')\boldsymbol{\nabla}'\psi(\mathbf{r}') - \psi(\mathbf{r}')\boldsymbol{\nabla}'\phi(\mathbf{r}')].$$
(5.1)

We take the function $\phi(\mathbf{r}')$ to be a solution of Poisson's equation

$$\boldsymbol{\nabla}'^2\phi(\mathbf{r}') = -4\pi\rho(\mathbf{r}').$$
(5.2)

We pick the other function ψ, designated as a **Green's Function** $G(\mathbf{r},\mathbf{r}')$, to be a function of two variables, satisfying

$$\boldsymbol{\nabla}'^2 G(\mathbf{r},\mathbf{r}') = -4\pi\delta(\mathbf{r}-\mathbf{r}').$$
(5.3)

Now Green's second theorem can be written as

$$\int_V [\phi(\mathbf{r}')\boldsymbol{\nabla}'^2 G(\mathbf{r},\mathbf{r}') - G(\mathbf{r},\mathbf{r}')\boldsymbol{\nabla}'^2\phi(\mathbf{r}')]d^3r' =$$
$$\oint_S d\mathbf{A}'\cdot[\phi(\mathbf{r}')\boldsymbol{\nabla}'G(\mathbf{r},\mathbf{r}') - G(\mathbf{r},\mathbf{r}')\boldsymbol{\nabla}'\phi(\mathbf{r}')],$$
(5.4)

where we integrate over $\mathbf{r}'$ keeping $\mathbf{r}$ fixed. Using Eqs. (5.2) and (5.3) in Eq. (5.4) results in

$$\phi(\mathbf{r}) = \int_V G(\mathbf{r},\mathbf{r}')\rho(\mathbf{r}')d^3r' - \frac{1}{4\pi}\oint_S d\mathbf{A}'\cdot[\phi(\mathbf{r}')\boldsymbol{\nabla}'G(\mathbf{r},\mathbf{r}') - G(\mathbf{r},\mathbf{r}')\boldsymbol{\nabla}'\phi(\mathbf{r}')].$$ (5.5)

This gives the solution to Poisson's equation in terms of a volume integral of the Green's function, and surface integrals of the Green's function and its normal derivative. In order to complete the solution to Poisson's equation, it is necessary to apply boundary conditions to be able to complete the surface integration. The surface, S, to be integrated over entails all surfaces bounding the volume V. If the volume is infinite, this includes a mathematical surface of infinite radius. For a multiplied connected volume (like Swiss cheese), the surface of each hole would be included.

5.2 Green's Function Solution of Poisson's Equation

We consider two relatively simple boundary conditions on the surface of the volume of integration that make the surface integrals in Eq. (5.5) tractable. If ϕ is given on any surface (**Dirichlet boundary condition**), then setting $G(\mathbf{r}, \mathbf{r}') = 0$ for $\mathbf{r}'$ on that surface will remove the second surface integral in Eq. (5.5) over the unknown normal derivative of ϕ. If it is the normal derivative of ϕ that is given on any surface (**Neumann boundary condition**), then setting the normal derivative with respect to $\mathbf{r}'$ of $G(\mathbf{r}, \mathbf{r}')$ equal to zero for $\mathbf{r}'$ on that surface will remove the first surface integral over the unknown surface potential ϕ. In either case, Eq. (5.5) will be the Green's function solution of Poisson's equation with the given boundary conditions. The general rule is that the Green's function satisfy the homogeneous boundary condition (Dirichlet or Neumann) corresponding to the inhomogeneous boundary condition satisfied by the potential.

5.2.1 Dirichlet Boundary Condition

For the Dirichlet boundary condition, the potential ϕ is specified on surfaces S that are bounding surfaces of the volume V for the volume integral in Eq. (5.5). For this case, the Green's function should be zero on the bounding surfaces, so that only the first surface integral in Eq. (5.5), where ϕ is known, enters. Then the Green's function solution to Poisson's equation with the potential specified on all bounding surfaces can be written as

$$\phi(\mathbf{r}) = \int_V G(\mathbf{r}, \mathbf{r}')\rho(\mathbf{r}')d^3r' - \frac{1}{4\pi}\oint_S d\mathbf{A}' \cdot [\phi(\mathbf{r}')\boldsymbol{\nabla}'G(\mathbf{r}, \mathbf{r}')]. \qquad (5.6)$$

Using the Green's function, the solution to Poisson's equation can be written down in two steps. The first step is immediate and straightforward. The solution is simply written down as given in terms of the Green's function by Eq. (5.6). The second, more difficult, step is to find the Green's function by using its two defining properties:

$$\text{I.} \qquad \nabla'^2 G(\mathbf{r}, \mathbf{r}') = -4\pi\delta(\mathbf{r} - \mathbf{r}'), \qquad (5.7)$$
$$\text{II.} \qquad G(\mathbf{r}, \mathbf{r}'_\mathbf{S}) = 0, \qquad (5.8)$$

where the notation $\mathbf{r}'_\mathbf{S}$ means that $\mathbf{r}'$ is on one of the bounding surfaces.

It is helpful to put the requirements for the Dirichlet Green's function into words:

> For the solution of Poisson's equation for an arbitrary charge distribution with the potential specified on all boundaries, the Green's function $G(\mathbf{r}, \mathbf{r}')$ is the potential at $\mathbf{r}'$ due to a unit point charge at $\mathbf{r}$ with all surfaces acting as grounded conductors.

Note that, even if the original surfaces are not conductors (they can't be conductors if the specified potential on them is not constant), the Green's function is the point charge solution found as if all the surfaces were grounded conductors.

We have discussed above the solution of Poisson's equation for the electric field. The Green's function method would apply in the same way for the gravitational field, or for a heat flow problem where the temperature acts as the potential and heat sources as the charge distribution. For the heat flow case, the temperature on the surface of the volume would be the Dirichlet boundary condition.

5.2.2   Surface Green's Function

For solving Laplace's equation, only the surface integral in Eq. (5.6) enters, and the solution can be written as

$$\phi(\mathbf{r}) = -\frac{1}{4\pi} \oint_S d\mathbf{A}' \cdot [\phi(\mathbf{r}')\nabla'G(\mathbf{r}, \mathbf{r}')]. \qquad (5.9)$$

We introduce a surface Green's function, defined by

$$g(\mathbf{r}, \mathbf{r}'_\mathbf{S}) = -\frac{1}{4\pi}\hat{\mathbf{n}}' \cdot \nabla'G(\mathbf{r}, \mathbf{r}')|_{\mathbf{r}'=\mathbf{r}'_\mathbf{S}}, \qquad (5.10)$$

where the unit vector $\hat{\mathbf{n}}'$ is directed out of the surface. Then, the Green's function solution to Laplace's equation can be written as

$$\phi(\mathbf{r}) = \oint_S dA'\, g(\mathbf{r}, \mathbf{r}'_\mathbf{S})\phi(\mathbf{r}'). \qquad (5.11)$$

The surface Green's function is given by Eq. (5.10), but it can also be found directly by considering what is required in Eq. (5.11). If the surface Green's function satisfies the two properties

$$\text{I.} \qquad \nabla^2 g(\mathbf{r}, \mathbf{r}_\mathbf{s}') = 0 \tag{5.12}$$

$$\text{II.} \qquad g(\mathbf{r}_\mathbf{s}, \mathbf{r}_\mathbf{s}') = \delta(\mathbf{r}_\mathbf{s} - \mathbf{r}_\mathbf{s}'), \tag{5.13}$$

the potential given by Eq. (5.11) will automatically satisfy Laplace's equation, and the Dirichlet boundary condition on the surface.

5.2.3   Neumann Boundary Condition

For Neumann boundary conditions, the normal derivative $\hat{\mathbf{n}} \cdot \nabla \phi$ of the potential at the surface is specified. In general, boundary conditions can be Neumann on some surfaces and Dirichlet on others. Here, we will consider only purely Neumann conditions, which require several added features.

If the potential is a solution to Poisson's equation in a volume V with Neumann boundary conditions on the surface, then the specified boundary values for the potential $\phi(\mathbf{r})$ must satisfy the constraint

$$\int_S d\mathbf{A} \cdot \nabla \phi = -4\pi \int_V \rho(\mathbf{r}) d^3 r. \tag{5.14}$$

This follows from applying the divergence theorem to Poisson's equation, written in the form

$$\nabla \cdot (\nabla \phi) = -4\pi \rho. \tag{5.15}$$

The Neumann Green's function has to satisfy the two conditions

$$\text{I.} \qquad \nabla'^2 G\mathbf{r}, \mathbf{r}') = -4\pi \delta(\mathbf{r} - \mathbf{r}') + \frac{4\pi}{V}, \tag{5.16}$$

$$\text{II.} \qquad \hat{\mathbf{n}}' \cdot \nabla' G(\mathbf{r}, \mathbf{r}')|_{\mathbf{r}'=\mathbf{r}_\mathbf{s}'} = 0. \tag{5.17}$$

The addition of the term $4\pi/V$ in Eq. (5.16) is required to satisfy the constraint

$$\int_S d\mathbf{A}' \cdot \nabla' G(\mathbf{r}, \mathbf{r}') = 0. \tag{5.18}$$

This constraint follows trivially because the integrand is zero to satisfy condition **II** for the Neumann Green's function. Without the added term, application of the divergence theorem to Eq. (5.16) would result in the surface integral of Eq. (5.18) not equaling zero.

Once the Green's function has been found, the solution of Poisson's equation is given by the integrals

$$\phi(\mathbf{r}) = \langle \phi \rangle + \int_V G(\mathbf{r}, \mathbf{r}') \rho(\mathbf{r}') d^3 r' + \frac{1}{4\pi} \oint_S d\mathbf{A}' \cdot [G(\mathbf{r}, \mathbf{r}') \nabla' \phi(\mathbf{r}')]. \tag{5.19}$$

The term $\langle\phi\rangle$, the average of the potential, comes from the term $4\pi/V$ in Eq. (5.16). It is an arbitrary constant that can be set to zero.

In all of the above, we have shown how Green's functions can be used to give integral solutions to Poisson's and Laplace's equations. Of course, actually finding the Green's function for a particular situation can be a difficult problem, but that is beyond the scope of our treatment of the vector calculus part of the problem.

5.3   Problems

1. (a) Show that the function

$$G(\pmb{\rho}, z; \pmb{\rho}', z') \;=\; \frac{1}{|\pmb{\rho} - \pmb{\rho}' + (z - z')\hat{\mathbf{k}}|} - \frac{1}{|\pmb{\rho} - \pmb{\rho}' - (z + z')\hat{\mathbf{k}}|}$$

defined in the half plane $z, z' \geq 0$, satisfies the two criteria for a Dirichlet Green's function. The vectors $\pmb{\rho}$ and $\pmb{\rho}'$ are two-dimensional vectors in the $z, z' = 0$ plane.

   (b) Find the surface Green's function for this Green's function.

   (c) Use the surface Green's function to find the potential along the positive z axis ($\rho = 0$) for the boundary condition

$$\begin{aligned}
\phi(\pmb{\rho}', 0) \;&=\; V. \quad \rho' < a \\
&=\; 0, \quad \rho' > a.
\end{aligned} \tag{5.20}$$

2. Show that the function

$$G(\mathbf{r}, \mathbf{r}') \;=\; \frac{1}{|\mathbf{r}' - \mathbf{r}|} - \frac{(R/r)}{|\mathbf{r}' - (R/r)^2 \mathbf{r}|}, \quad r, r' \leq R,$$

defined inside a sphere of radius R, satisfies the two criteria for a Dirichlet Green's function.

Chapter 6

General Coordinate Systems

A general coordinate system with three variables q_1, q_2, q_3 can be defined by relating the general coordinates to the Cartesian coordinates $x_1(= x)$, $x_2(= y)$, $x_3(= z)$. The defining equations can be taken as the three functions $x_i(q_1, q_2, q_3)$.

We would like to relate the unit basis vectors $\hat{\mathbf{q}}_i$ in the general system to the Cartesian unit vectors $\hat{\mathbf{n}}_1(= \hat{\mathbf{i}})$, $\hat{\mathbf{n}}_2(= \hat{\mathbf{j}})$, $\hat{\mathbf{n}}_3(= \hat{\mathbf{k}})$. The differential of the position vector $\mathbf{r}$ can be written as

$$
\begin{aligned}
\mathbf{dr} &= \sum_k \hat{\mathbf{n}}_k dx_k \\
&= \sum_k \hat{\mathbf{n}}_k \sum_i \frac{\partial x_k}{\partial q_i} dq_i \\
&= \sum_i \left[\sum_k \frac{\partial x_k}{\partial q_i} \hat{\mathbf{n}}_k \right] dq_i.
\end{aligned}
\tag{6.1}
$$

A unit basis vector $\hat{\mathbf{q}}_i$ in the general system is defined as that direction in which only the coordinate q_i changes. It can be seen from the last step in Eq. (6.1) that the quantity in square brackets is in the direction of $\hat{\mathbf{q}}_i$, but it may not have unit magnitude. We introduce **metric coefficients** h_i such that $h_i dq_i$ is the distance moved when the coordinate q_i changes by the differential amount dq_i, with the other general coordinates fixed. Then we can write

$$
\mathbf{dr} = \sum_i h_i \hat{\mathbf{q}}_i dq_i
\tag{6.2}
$$

where

$$
h_i \hat{\mathbf{q}}_i = \sum_k \frac{\partial x_k}{\partial q_i} \hat{\mathbf{n}}_k.
\tag{6.3}
$$

We will restrict our considerations to orthogonal coordinate systems, defined by the condition

$$
\hat{\mathbf{q}}_i \cdot \hat{\mathbf{q}}_j = \delta_{ij}.
\tag{6.4}
$$

45

This gives the orthogonality condition that

$$\sum_k \frac{\partial x_k}{\partial q_i} \frac{\partial x_k}{\partial q_j} = h_i h_j \delta_{ij}. \tag{6.5}$$

For $j = i$, Eq. (6.5) gives the magnitude of the metric coefficient h_i as

$$h_i = \sqrt{\sum_k \left(\frac{\partial x_k}{\partial q_i}\right)^2}. \tag{6.6}$$

It is also convenient to have a right-handed coordinate system, which is defined by

$$\hat{\mathbf{q}}_1 \times \hat{\mathbf{q}}_2 = \hat{\mathbf{q}}_3, \quad \text{and cyclic.} \tag{6.7}$$

6.1   Vector Differential Operators

We can now use the definitions of the vector differential operators to find their forms in a general orthogonal coordinate system. The gradient is defined by Eq. (1.2):

$$d\phi = \mathbf{dr} \cdot \boldsymbol{\nabla}\phi = \sum_i (\boldsymbol{\nabla}\phi)_i h_i dq_i, \tag{6.8}$$

where we have used the relation $\mathbf{dr} = \sum_i h_i \hat{\mathbf{n}}_i dq_i$ in the second step above. We can also expand the differential $d\phi$ as

$$d\phi = \sum_i \left(\frac{\partial \phi}{\partial q_i}\right) dq_i. \tag{6.9}$$

Comparing Eqs. (6.8) and (6.9), and using the fact that the displacements dq_i are arbitrary, we see that

$$(\boldsymbol{\nabla}\phi)_i = \frac{1}{h_i}\frac{\partial \phi}{\partial q_i}. \tag{6.10}$$

The definition of the divergence of a vector given by Eq. (1.17) is

$$\boldsymbol{\nabla} \cdot \mathbf{E} = \lim_{V \to 0} \frac{1}{V} \oint_S \mathbf{E} \cdot \mathbf{dA}. \tag{6.11}$$

The derivation following Eq. (1.17) for the form of the divergence in Cartesian coordinates can be extended to a general coordinate system. For general coordinates, the volume V to be divided by in Eq. (6.11) becomes

$$V = h_1 \Delta q_1 h_2 \Delta q_2 h_3 \Delta q_3. \tag{6.12}$$

The surface integrals are also modified by the h_is, with the area

$$dxdy \to h_1 h_2 dq_1 dq_2, \quad \text{and cyclic.} \tag{6.13}$$

Then, for general coordinates, Eq. (1.22) becomes

$$
\begin{aligned}
I+II &= \lim_{\Delta q_i \to 0} \frac{[(E_1 h_2 h_3)(q_1 + \Delta q_1, q_2, q_3) - (E_1 h_2 h_3)(q_1, q_2, q_3)]\Delta q_2 \Delta q_3}{h_1 h_2 h_3 \Delta q_1 \Delta q_2 \Delta q_3} \\
&= \lim_{\Delta q_1 \to 0} \frac{[(E_1 h_2 h_3)(q_1 + \Delta q_1, q_2, q_3) - (E_1 h_2 h_3)(q_1, q_2, q_3)]}{h_1 h_2 h_3 \Delta q_1} \\
&= \frac{1}{h_1 h_2 h_3} \frac{\partial}{\partial q_1} (E_1 h_2 h_3).
\end{aligned}
\tag{6.14}
$$

The integrals over faces III, IV, V, and VI are done similarly, so the total divergence for general coordinates can be written as

$$\mathbf{\nabla \cdot E} = \frac{1}{h_1 h_2 h_3} \left[\frac{\partial}{\partial q_1} (E_1 h_2 h_3) + \text{cyclic} \right]. \tag{6.15}$$

The derivation of this general result can be used as a mnemonic for the form of the divergence in general coordinates. That is, the denominator $h_1 h_2 h_3$ can be remembered as the division by the volume element, while the factor $h_2 h_3$ multiplying E_1 can be recognized as the area of the perpendicular surface element. Then the other two terms just come by cyclic substitution.

The Laplacian operator is the divergence of a gradient, so Eqs. (6.10) and (6.15) can be combined to give

$$\nabla^2 \phi = \frac{1}{h_1 h_2 h_3} \left\{ \frac{\partial}{\partial q_1} \left[\left(\frac{h_2 h_3}{h_1} \right) \frac{\partial \phi}{\partial q_1} \right] + \text{cyclic} \right\}. \tag{6.16}$$

The contribution to the z component of the curl of a vector given by the definition of curl in Eq. (1.26) can be extended to a general coordinate system. In Cartesian coordinates, the contribution to Eq. (1.26) for the z component of curl from sides I and II in Fig. 1.3 was

$$I + II = \lim_{\Delta x, \Delta y \to 0} \frac{[E_y(x + \Delta x, y, z)\Delta y - E_y(x, y, z)\Delta y]}{\Delta x \Delta y}. \tag{6.17}$$

For general coordinates q_i, this becomes $I + II =$

$$\lim_{\Delta q_1, \Delta q_2 \to 0} \frac{[E_2(q_1 + \Delta q_1, q_2, q_3)h_2(q_1 + \Delta q_1, q_2, q_3)\Delta q_2 - E_2(q_1, q_2, q_3)h_1(q_1, q_2, q_3)\Delta q_2]}{h_1 \Delta q_1 h_2 \Delta q_2}.$$

$$\tag{6.18}$$

Here, the actual distance corresponding to each coordinate displacement Δq_i has been given by $h_i \Delta q_i$. Now, taking the limit $\Delta q_1, \Delta q_2 \to 0$, gives the result

$$I + II = \frac{1}{h_1 h_2} \frac{\partial (h_2 E_2)}{\partial q_1}. \tag{6.19}$$

The procedure for the sides III and IV is the same, with a sign change due to the opposite direction of integration, leading to the general expression for the curl in general coordinates:

$$(\mathbf{curl\,E})_3 = \frac{1}{h_1 h_2} \left[\frac{\partial}{\partial q_1}(h_2 E_2) - \frac{\partial}{\partial q_2}(h_1 E_1) \right], \text{ and cyclic.} \tag{6.20}$$

The first term in Eq. (6.20) is in the cyclic order 1,2, and is positive. Then second term has a minus sign, and interchanges the indices 1 and 2. The other two components of **curl** are given by cyclic substitution. The $h_1 h_2$ in the denominator can be remembered as the division by the area perpendicular to the 3 direction, and the factor $h_2 E_2$ can be seen as coming from the line integral of **E·dr** in the 2 direction in the definition of curl in Eq. (1.26).

6.2 Spherical Coordinates

In this section, we apply the relations derived above for a general coordinate system to the important physical case of spherical coordinates.

Spherical coordinates, shown in Fig. 6.1, are defined by

$$\begin{aligned} q_1 &= r, & \text{the radial distance,} \tag{6.21}\\ q_2 &= \theta, & \text{the polar angle,} \tag{6.22}\\ q_3 &= \phi, & \text{the azimuthal angle.} \tag{6.23}\end{aligned}$$

They are related to the Cartesian coordinates by

$$\begin{aligned} x &= r \sin\theta \cos\phi \tag{6.24}\\ y &= r \sin\theta \sin\phi \tag{6.25}\\ z &= r \cos\theta, \tag{6.26}\end{aligned}$$

as can be seen from Fig. 6.1.

The spherical coordinate system is an orthogonal system, which can be verified by substitution of Eqs. (6.24) - (6.26) into the orthogonality condition of Eq. (6.5). The metric coefficients h_i can be calculated from Eq. (6.6), or by observing from Fig. 6.1 that

$$\mathbf{dr} = dr\hat{\mathbf{r}} + r d\theta \hat{\boldsymbol{\theta}} + r \sin\theta d\phi \hat{\boldsymbol{\phi}}. \tag{6.27}$$

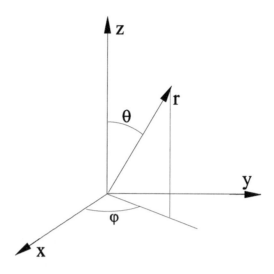

Figure 6.1: Spherical coordinates.

Either way gives

$$h_1 = h_r = 1 \tag{6.28}$$
$$h_2 = h_\theta = r \tag{6.29}$$
$$h_3 = h_\phi = r\sin\theta. \tag{6.30}$$

The spherical unit vectors $\hat{\mathbf{r}}, \hat{\boldsymbol{\theta}}, \hat{\boldsymbol{\phi}}$ can be related to the Cartesian unit vectors $\hat{\mathbf{i}}, \hat{\mathbf{j}}, \hat{\mathbf{k}}$ using Eq. (6.3) or by observation from Fig. 6.1, giving

$$\hat{\mathbf{r}} = \sin\theta\cos\phi\hat{\mathbf{i}} + \sin\theta\sin\phi\hat{\mathbf{j}} + \cos\theta\hat{\mathbf{k}}, \tag{6.31}$$
$$\hat{\boldsymbol{\theta}} = \cos\theta\cos\phi\hat{\mathbf{i}} + \cos\theta\sin\phi\hat{\mathbf{j}} - \sin\theta\hat{\mathbf{k}}, \tag{6.32}$$
$$\hat{\boldsymbol{\phi}} = -\sin\phi\hat{\mathbf{i}} + \cos\phi\hat{\mathbf{j}}. \tag{6.33}$$

With the spherical h_i known, we can substitute into the general equations for grad, div, curl, and the Laplacian to get

$$\boldsymbol{\nabla}\psi = \hat{\mathbf{r}}(\partial_r\psi) + \frac{\hat{\boldsymbol{\theta}}(\partial_\theta\psi)}{r} + \frac{\hat{\boldsymbol{\phi}}(\partial_\phi\psi)}{r\sin\theta}, \tag{6.34}$$

$$\boldsymbol{\nabla}\cdot\mathbf{E} = \frac{1}{r^2\sin\theta}[\partial_r(r^2\sin\theta E_r) + \partial_\theta(r\sin\theta E_\theta) + \partial_\phi(rE_\phi)]$$

$$= \frac{\partial_r(r^2 E_r)}{r^2} + \frac{\partial_\theta(\sin\theta E_\theta)}{r\sin\theta} + \frac{(\partial_\phi E_\phi)}{r\sin\theta}, \tag{6.35}$$

$$\boldsymbol{\nabla}\times\mathbf{E} = \frac{\hat{\mathbf{r}}[\partial_\theta(r\sin\theta E_\phi) - \partial_\phi(rE_\theta)]}{r^2\sin\theta} + \frac{\hat{\boldsymbol{\theta}}[(\partial_\phi E_r) - \partial_r(r\sin\theta E_\phi)]}{r\sin\theta}$$

$$+\frac{\hat{\phi}[\partial_r(rE_\theta) - \partial_\theta E_r]}{r}, \tag{6.36}$$

$$\nabla^2\psi = \frac{[\partial_r(r^2\partial_r\psi)]}{r^2} + \frac{[\partial_\theta(\sin\theta\partial_\theta\psi)]}{r^2\sin\theta} + \frac{[\partial_\phi^2\psi]}{r^2\sin^2\theta}. \tag{6.37}$$

For spherical coordinates, we have designated the potential as $\psi(r, \theta, \phi)$ to avoid confusion between the potential and the ϕ coordinate.

6.3 Cylindrical Coordinates

Cylindrical coordinates, shown in Fig. 6.2, are defined by

$$\begin{aligned} q_1 &= r, \quad &\text{the distance from the symmetry axis,} \tag{6.38} \\ q_2 &= \theta, \quad &\text{the angle about the symmetry axis,} \tag{6.39} \\ q_3 &= z, \quad &\text{the distance along the symmetry axis.} \tag{6.40} \end{aligned}$$

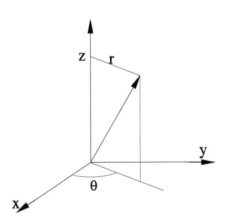

Figure 6.2: Cylindrical coordinates.

They are related to the Cartesian coordinates by

$$\begin{aligned} x &= r\cos\theta \tag{6.41} \\ y &= r\sin\theta \tag{6.42} \\ z &= z, \tag{6.43} \end{aligned}$$

as can be seen from Fig. 6.2 . Note that the radial coordinate r and the angle θ in cylindrical coordinates are different than the r and θ used in spherical coordinates.

The cylindrical coordinate system is an orthogonal system, as can be verified by substitution of Eqs. (6.41) - (6.43) into the orthogonality condition of Eq. (6.5). The metric coefficients h_i can be calculated from Eq. (6.6) or by observation from Fig. 6.2:

$$h_1 = h_r = 1 \tag{6.44}$$
$$h_2 = h_\theta = r \tag{6.45}$$
$$h_3 = h_z = 1. \tag{6.46}$$

The cylindrical unit vectors $\hat{\mathbf{r}}, \hat{\boldsymbol{\theta}}, \hat{\mathbf{z}}$ can be related to the Cartesian unit vectors $\hat{\mathbf{i}}$, $\hat{\mathbf{j}}$, $\hat{\mathbf{k}}$ using Eq. (6.3), or by observation from Fig. 6.2, giving

$$\hat{\mathbf{r}} = \cos\theta\hat{\mathbf{i}} + \sin\theta\hat{\mathbf{j}}, \tag{6.47}$$
$$\hat{\boldsymbol{\theta}} = -\sin\theta\hat{\mathbf{i}} + \cos\theta\hat{\mathbf{j}}, \tag{6.48}$$
$$\hat{\mathbf{z}} = \hat{\mathbf{k}}. \tag{6.49}$$

With the cylindrical h_is known, we can substitute into the general equations for grad, div, curl, and the Laplacian to get

$$\boldsymbol{\nabla}\psi = \hat{\mathbf{r}}(\partial_r\psi) + \hat{\boldsymbol{\theta}}(\partial_\theta\psi)/r + \hat{\mathbf{z}}(\partial_z\psi), \tag{6.50}$$
$$\boldsymbol{\nabla}{\cdot}\mathbf{E} = (1/r)[\partial_r(rE_r) + (\partial_\theta E_\theta)] + \partial_z(E_z), \tag{6.51}$$
$$\boldsymbol{\nabla}{\times}\mathbf{E} = \hat{\mathbf{r}}[(1/r)(\partial_\theta E_z) - (\partial_z E_\theta)] + \hat{\boldsymbol{\theta}}[\partial_z E_r - (\partial_r E_z)]$$
$$+ (\hat{\mathbf{z}}/r)[\partial_r(rE_\theta) - (\partial_\theta E_r)], \tag{6.52}$$
$$\boldsymbol{\nabla}^2\psi = (1/r)[\partial_r(r\partial_r\psi)] + (1/r^2)[\partial_\theta^2\psi)] + \partial_z^2\psi. \tag{6.53}$$

6.4   Problems

1. (a) Write the potential of an electric dipole **p** in Cartesian coordinates.

   (b) Find **E** of the dipole for $r>0$ by taking the negative gradient in Cartesian coordinates. Show that this equals the result of writing the vector expression for **E** in Cartesian coordinates.

   (c) Calculate the curl and the divergence of **E** of the dipole in Cartesian coordinates for $r>0$. (Each should come out zero.)

2. (a) Verify algebraically that spherical coordinates form an orthogonal coordinate system.

   (b) Calculate the metric coefficients h_i for spherical coordinates algebraically.

3. (a) Write the potential of an electric dipole **p** (pointing in the z direction) in spherical coordinates.

   (b) Find **E** of the dipole for $r>0$ by taking the negative gradient in spherical coordinates. Show that this equals the result of writing the vector expression for **E** in spherical coordinates.

   (c) Calculate the curl and the divergence of **E** of the dipole in spherical coordinates for $r>0$. (Each should come out zero.)

4. (a) Write the potential of an electric dipole **p** (pointing in a general direction) in spherical coordinates.

   (b) Find **E** of the dipole for $r>0$ by taking the negative gradient in spherical coordinates. Show that this equals the result of writing the vector expression for **E** in spherical coordinates.

   (c) Calculate the curl and the divergence of **E** of the dipole in spherical coordinates for $r>0$. (Each should come out zero.)

5. (a) Verify algebraically that cylindrical coordinates form an orthogonal coordinate system.

   (b) Calculate the metric coefficients h_i for cylindrical coordinates algebraically.

6. (a) Write the potential of a dipole **p** (in the z direction) in cylindrical coordinates.

   (b) Find **E** of the dipole by taking the negative gradient in cylindrical coordinates. Show that this equals the result of writing the vector expression for **E** in cylindrical coordinates.

7. (a) Write the potential of an electric dipole **p** (in a general direction) in cylindrical coordinates.

  (b) Find **E** of the dipole by taking the negative gradient in cylindrical coordinates. Show that this equals the result of writing the vector expression for **E** in cylindrical coordinates.

Chapter 7

Dyadics

Dyadics are an extension of vector notation that are useful in treating quadrupoles, the tensor of inertia, and general non-isotropic relations between vectors. They are more commonly called tensors rather than dyadics because they transform like a second rank tensor under coordinate transformations. Our treatment below of dyadics as extensions of vectors is more useful in calculations than tensor notation, which is usually developed in the Cartesian coordinate system.

7.1   Dyad

We first give the definition of a **dyad**, which is the direct product of two vectors written without a dot or a cross as $\mathbf{AB}$. The meaning of $\mathbf{AB}$ is given by the following operations with a third vector $\mathbf{C}$:

$$\mathbf{AB}\cdot\mathbf{C} = \mathbf{A}(\mathbf{B}\cdot\mathbf{C}), \qquad \mathbf{C}\cdot\mathbf{AB} = (\mathbf{C}\cdot\mathbf{A})\mathbf{B} \qquad (7.1)$$

$$\mathbf{AB}\times\mathbf{C} = \mathbf{A}(\mathbf{B}\times\mathbf{C}), \qquad \mathbf{C}\times\mathbf{AB} = (\mathbf{C}\times\mathbf{A})\mathbf{B}, \qquad (7.2)$$

with $\mathbf{A}(\mathbf{B}\times\mathbf{C})$ and $(\mathbf{C}\times\mathbf{A})\mathbf{B}$ being new dyads.

There is also a double dot operation that has the following meaning:

$$\mathbf{AB} : \mathbf{CD} = \mathbf{A}\cdot(\mathbf{B}\cdot\mathbf{C})\mathbf{D} = (\mathbf{A}\cdot\mathbf{D})(\mathbf{B}\cdot\mathbf{C}). \qquad (7.3)$$

That is, we first dot $\mathbf{B}$ with $\mathbf{CD}$ resulting in the vector $(\mathbf{B}\cdot\mathbf{C})\mathbf{D}$, and then $\mathbf{A}$ is dotted into $(\mathbf{B}\cdot\mathbf{C})\mathbf{D}$.

7.2   Dyadic

A **dyadic** is a more general object, which we represent as $[\mathbf{D}]$. The dyadic can be represented by a matrix, with nine independent matrix elements defined by

$$D_{ij} = \hat{\mathbf{n}}_i\cdot[\mathbf{D}]\cdot\hat{\mathbf{n}}_j, \qquad (7.4)$$

where $\hat{\mathbf{n}}_1$, $\hat{\mathbf{n}}_2$, $\hat{\mathbf{n}}_3$ are the three basic unit vectors. The vectors $\hat{\mathbf{n}}_i$ can be basic unit vectors in any coordinate system, but we will consider only orthogonal systems.

A dyadic can be expanded as sums of dyads. The standard expansion is in terms of the basic unit vectors:

$$[\mathbf{D}] = \Sigma_{ij} D_{ij} \hat{\mathbf{n}}_i \hat{\mathbf{n}}_j. \tag{7.5}$$

If the coordinate axes were not orthogonal, the unit vectors in Eq. (7.4) and Eq. (7.5) would not be the same, and the two sets of D_{ij} would also be different. However, for orthogonal coordinate systems the $\hat{\mathbf{n}}_i$ and D_{ij} are the same in each equation.

An important dyadic is the **unit dyadic** $\hat{\hat{\mathbf{n}}}$, defined by its dot product with a general vector $\mathbf{A}$:

$$\hat{\hat{\mathbf{n}}} \cdot \mathbf{A} = \mathbf{A} \cdot \hat{\hat{\mathbf{n}}} = \mathbf{A}. \tag{7.6}$$

Expanded in the basic unit vectors, the unit dyadic is given by

$$\hat{\hat{\mathbf{n}}} = \Sigma_i \hat{\mathbf{n}}_i \hat{\mathbf{n}}_i. \tag{7.7}$$

Its matrix representation is the unit matrix with elements given by

$$[\hat{\hat{\mathbf{n}}}]_{ij} = \delta_{ij}. \tag{7.8}$$

A useful identity for the unit dyadic is that

$$\boldsymbol{\nabla}\mathbf{r} = \hat{\hat{\mathbf{n}}}. \tag{7.9}$$

This follows from the two relations

$$\mathbf{p} \cdot [\boldsymbol{\nabla}\mathbf{r}] = (\mathbf{p} \cdot \boldsymbol{\nabla})\mathbf{r} = \mathbf{p} \quad \text{and} \quad [\boldsymbol{\nabla}\mathbf{r}] \cdot \mathbf{p} = \boldsymbol{\nabla}(\mathbf{r} \cdot \mathbf{p})|_{\mathbf{p}_{\text{constant}}} = \mathbf{p}. \tag{7.10}$$

The use of dyadics will be made clear in the derivations below.

7.3 Three-dimensional Taylor Expansion

One use of dyadics is in the three-dimensional **Taylor expansion**, which in dyadic notation is given by

$$\begin{aligned} \phi(\mathbf{r}) = \ & \phi(0) + \mathbf{r} \cdot [\boldsymbol{\nabla}\phi(\mathbf{r})]_{\mathbf{r}=0} + \frac{1}{2}\mathbf{r}\mathbf{r} : [\boldsymbol{\nabla}\boldsymbol{\nabla}\phi(\mathbf{r})]_{\mathbf{r}=0} \\ & + \frac{1}{6}\mathbf{r}\mathbf{r}\mathbf{r} \vdots [\boldsymbol{\nabla}\boldsymbol{\nabla}\boldsymbol{\nabla}\phi(\mathbf{r})]_{\mathbf{r}=0} + \dots \end{aligned} \tag{7.11}$$

7.4 Multipole Expansion

The Taylor expansion can be used to generate the **multipole expansion** of the Coulomb integral for the potential,

$$\phi(\mathbf{r}) = \int \frac{\rho(\mathbf{r}')d^3r'}{|\mathbf{r} - \mathbf{r}'|}. \tag{7.12}$$

For a charge distribution that is restricted in extent, r can be chosen to always be greater than the largest r' in the integral, so that $r'/r < 1$. Then the Taylor expansion will converge, and integrating each term in the expansion separately will give a multipole expansion of the potential in powers of $1/r$.

We look at the first three terms in the expansion of the denominator of the integrand, treated as a function of $\mathbf{r}'$ with $\mathbf{r}$ considered a constant vector:

$$\frac{1}{|\mathbf{r} - \mathbf{r}'|} = \left[\frac{1}{|\mathbf{r} - \mathbf{r}'|}\right]_{\mathbf{r}'=0} + \mathbf{r}' \cdot \left[\mathbf{\nabla}'\left(\frac{1}{|\mathbf{r} - \mathbf{r}'|}\right)\right]_{\mathbf{r}'=0} + \frac{1}{2}\mathbf{r}'\mathbf{r}' : \left[\mathbf{\nabla}'\mathbf{\nabla}'\left(\frac{1}{|\mathbf{r} - \mathbf{r}'|}\right)\right]_{\mathbf{r}'=0}. \tag{7.13}$$

When put into the Coulomb integral, the first term of the Taylor expansion leads to the zeroth order potential

$$\phi_0(\mathbf{r}) = \int \frac{\rho(\mathbf{r}')d^3r'}{r} = \frac{q}{r}. \tag{7.14}$$

This is the potential due to a point charge whose magnitude is equal to the net charge of the charge distribution. It is called the **monopole potential** because it is equivalent to the potential of a single point charge.

7.5 Dipole Moment

The second term in the Taylor expansion is

$$\mathbf{r}' \cdot \left[\mathbf{\nabla}'\left(\frac{1}{|\mathbf{r} - \mathbf{r}'|}\right)\right]_{\mathbf{r}'=0} = \mathbf{r}' \cdot \left[\frac{(\mathbf{r} - \mathbf{r}')}{|\mathbf{r} - \mathbf{r}'|^3}\right]_{\mathbf{r}'=0} = \frac{\mathbf{r}\cdot\mathbf{r}'}{r^3}. \tag{7.15}$$

Putting this term into the Coulomb integral leads to the first order potential

$$\phi_1(\mathbf{r}) = \frac{\mathbf{r}}{r^3} \cdot \int \mathbf{r}'\rho(\mathbf{r}')d^3r'. \tag{7.16}$$

The **electric dipole moment** of a charge distribution is defined by

$$\mathbf{p} = \int \mathbf{r}\rho(\mathbf{r})d^3r, \tag{7.17}$$

and then Eq. (7.16) can be written as

$$\phi_1(\mathbf{r}) = \frac{\mathbf{r}\cdot\mathbf{p}}{r^3}. \tag{7.18}$$

The potential term ϕ_1 is called the **dipole potential**.

The charge differential element $dq = \rho d^3 r$ used above is for a three-dimensional charge distribution within a volume. This can be generalized to one- or two-dimensional charge distributions as

$$dq = \lambda dl \quad \text{or} \quad dq = \sigma dA, \tag{7.19}$$

where λ is the charge per unit length, and σ the surface charge per unit area.

7.6   Quadrupole Dyadic

For the third term in the expansion, we must evaluate the dyadic

$$
\begin{aligned}
\nabla'\nabla'\left(\frac{1}{|\mathbf{r}-\mathbf{r}'|}\right) &= \nabla'\left(\frac{(\mathbf{r}-\mathbf{r}')}{|\mathbf{r}-\mathbf{r}'|^3}\right) \\
&= \frac{3(\mathbf{r}-\mathbf{r}')(\mathbf{r}-\mathbf{r}')}{|\mathbf{r}-\mathbf{r}'|^5} - \frac{\hat{\mathbf{n}}}{|\mathbf{r}-\mathbf{r}'|^3}.
\end{aligned}
\tag{7.20}
$$

Note that the operator ∇' acts only on the vector $\mathbf{r}'$, with $\mathbf{r}$ being considered a constant vector here. In the second term above, we have used the identity $\nabla\mathbf{r} = \hat{\mathbf{n}}$. We have not included a singular term in Eq. (7.20) because a quadrupole is generally an extended object, and nothing like a point quadrupole has been observed.

Setting $\mathbf{r}' = \mathbf{0}$ in Eq. (7.20), we get

$$\left[\nabla'\nabla'\left(\frac{1}{|\mathbf{r}-\mathbf{r}'|}\right)\right]_{\mathbf{r}'=0} = \frac{3\hat{\mathbf{r}}\hat{\mathbf{r}} - \hat{\mathbf{n}}}{r^3}. \tag{7.21}$$

Using this result in Eq. (7.13), the second order potential is given by

$$\phi_2(\mathbf{r}) = \left(\frac{3\hat{\mathbf{r}}\hat{\mathbf{r}} - \hat{\mathbf{n}}}{2r^3}\right) : \int \mathbf{r}'\mathbf{r}'dq'. \tag{7.22}$$

The dyadic $(3\hat{\mathbf{r}}\hat{\mathbf{r}} - \hat{\mathbf{n}})$ has the property that

$$(3\hat{\mathbf{r}}\hat{\mathbf{r}} - \hat{\mathbf{n}}) : \hat{\mathbf{n}} = 0. \tag{7.23}$$

This follows from the identity

$$\hat{\mathbf{n}} : \hat{\mathbf{n}} = 3, \tag{7.24}$$

that is the trace of the unit dyadic is 3. Using this result, we can modify the integrand in Eq. (7.22) by subtracting the dyadic $\frac{1}{3}r'^2\hat{\mathbf{n}}$, so the potential ϕ_2 can be written

$$\phi_2(\mathbf{r}) = \left(\frac{3\hat{\mathbf{r}}\hat{\mathbf{r}} - \hat{\mathbf{n}}}{2r^3}\right) \int (\mathbf{r}'\mathbf{r}' - \frac{1}{3}r'^2\hat{\mathbf{n}})dq'. \tag{7.25}$$

The **quadrupole dyadic** is defined as

$$[\mathbf{Q}] = \frac{1}{2} \int (3\mathbf{r}\mathbf{r} - r^2\hat{\mathbf{n}})dq'. \tag{7.26}$$

Then, the potential ϕ_2 can be written as

$$\phi_2(\mathbf{r}) = \frac{(3\hat{\mathbf{r}}\hat{\mathbf{r}} - \hat{\mathbf{n}}) : [\mathbf{Q}]}{3r^3}, \tag{7.27}$$

and is called the **quadrupole potential**.

From the property (7.23), it follows that

$$\hat{\mathbf{n}} : [\mathbf{Q}] = 0, \tag{7.28}$$

so Eq. (7.27) can also be written

$$\phi_2(\mathbf{r}) = \frac{\hat{\mathbf{r}}\hat{\mathbf{r}} : [\mathbf{Q}]}{r^3}. \tag{7.29}$$

Either Eq. (7.27) or Eq. (7.29) for the quadrupole potential can be used, depending on which is more convenient in any particular application.

In the above, we have used the Taylor expansion in the Coulomb integral to generate a multipole expansion of the potential in powers of $1/r$ given by

$$\begin{aligned} \phi(\mathbf{r}) &= \phi_0(\mathbf{r}) + \phi_1(\mathbf{r}) + \phi_2(\mathbf{r}) + O\left(\frac{1}{r^4}\right) \\ &= \frac{q}{r} + \frac{\hat{\mathbf{r}}\cdot\mathbf{p}}{r^2} + \frac{\hat{\mathbf{r}}\hat{\mathbf{r}} : [\mathbf{Q}]}{r^3} + O\left(\frac{1}{r^4}\right), \end{aligned} \tag{7.30}$$

with each order of potential in Eq. (7.30) coming from the corresponding term in Eq. (7.13). The terms in Eq. (7.30) are identified as the monopole, dipole, and quadrupole potentials, respectively.

The next term in the multipole expansion would be the octupole potential. But as can be seen from Eq. (7.11), it would involve the triple dot product of two three-dyadics. This would be exceedingly complicated, so multipoles higher than the quadrupole are usually treated by other means (such as spherical harmonics).

The quadrupole dyadic can be represented by a matrix with matrix elements D_{ij} given by Eq. (7.4). The quadrupole dyadic transforms under rotations like a second rank tensor. Since it is real and symmetric, it can be diagonalized by a rotation of coordinates, and can be represented in Cartesian coordinates by the diagonal matrix

$$[\mathbf{Q}] = \begin{bmatrix} Q_x & 0 & 0 \\ 0 & Q_y & 0 \\ 0 & 0 & Q_z \end{bmatrix}. \tag{7.31}$$

From the property given by Eq. (7.23), the quadrupole matrix has the trace constraint

$$Q_x + Q_y + Q_z = 0. \tag{7.32}$$

Very often the symmetry of the charge distribution results in two of the diagonal quadrupole elements being equal. These are usually chosen to be Q_x and Q_y, and the diagonal matrix representation of the quadrupole is written as

$$[\mathbf{Q}] = Q_0 \begin{bmatrix} -\frac{1}{2} & 0 & 0 \\ 0 & -\frac{1}{2} & 0 \\ 0 & 0 & 1 \end{bmatrix}. \tag{7.33}$$

The quantity Q_0 is referred to as **'the quadrupole moment'** of the charge distribution. It is given in Cartesian coordinates by

$$Q_0 = \frac{1}{2} \int (2z^2 - x^2 - y^2) dq, \tag{7.34}$$

and in spherical coordinates by

$$Q_0 = \frac{1}{2} \int (3 \cos^2 \theta - 1) r^2 dq. \tag{7.35}$$

We can see from Eq. (7.34) that Q_0 will generally be positive for an elongated charge distribution (American football) and negative for a squashed distribution (pancake).

The symmetric quadrupole can also be written in dyadic notation as

$$[\mathbf{Q}] = \frac{1}{2} Q_0 [3\hat{\mathbf{k}}\hat{\mathbf{k}} - \hat{\mathbf{n}}], \tag{7.36}$$

where $\hat{k}$ is the symmetry axis of the quadrupole. This dyadic form is useful for vector calculations.

Substituting the dyadic form of $[\mathbf{Q}]$ into Eq. (7.29), the potential for a symmetric quadrupole can be written as

$$\phi = \frac{Q_0}{2r^3} [3(\hat{\mathbf{k}} \cdot \hat{\mathbf{r}})^2 - 1]. \tag{7.37}$$

The electric field is the negative gradient of this potential, given by

$$\mathbf{E} = \frac{3Q_0}{2r^4} [5(\hat{\mathbf{k}} \cdot \hat{\mathbf{r}})^2 \hat{\mathbf{r}} - 2(\hat{\mathbf{k}} \cdot \hat{\mathbf{r}})\hat{\mathbf{k}} - \hat{\mathbf{r}}]. \tag{7.38}$$

7.7 Tensor (Dyadic) of Inertia

The angular momentum of an extended object is given by

$$\mathbf{L} = \int \mathbf{r} \times \mathbf{v}(\mathbf{r}) \rho(\mathbf{r}) d^3 r, \tag{7.39}$$

where $\rho(\mathbf{r})$ is the mass density in the object, and $\mathbf{v}(\mathbf{r})$ is the velocity of the matter within the object. If the object is undergoing rigid rotation with an angular velocity $\boldsymbol{\omega}$, the velocity of the matter in the object is given by

$$\mathbf{v}(\mathbf{r}) = \boldsymbol{\omega} \times \mathbf{r}, \tag{7.40}$$

and the angular momentum is given by

$$\begin{aligned} \mathbf{L} &= \int \mathbf{r} \times (\boldsymbol{\omega} \times \mathbf{r}) \rho(\mathbf{r}) d^3 r \\ &= \int [r^2 \boldsymbol{\omega} - \mathbf{r}(\mathbf{r} \cdot \boldsymbol{\omega})] \rho(\mathbf{r}) d^3 r. \end{aligned} \tag{7.41}$$

We define a **dyadic of inertia**[1]

$$[\mathbf{I}] = \int (r^2 \hat{\hat{\mathbf{n}}} - \mathbf{rr}) \rho(\mathbf{r}) d^3 r. \tag{7.42}$$

Then the angular momentum of the extended object can be written as

$$\mathbf{L} = [\mathbf{I}] \cdot \boldsymbol{\omega}. \tag{7.43}$$

The dyadic of inertia can be represented by a matrix with matrix elements I_{ij} given by Eq. (7.4). In Cartesian coordinates, these matrix elements are

$$[\mathbf{I}] = \begin{bmatrix} I_{xx} & I_{xy} & I_{xz} \\ I_{yx} & I_{yy} & I_{yz} \\ I_{zx} & I_{zy} & I_{zz} \end{bmatrix}. \tag{7.44}$$

The diagonal elements, called **moments of inertia**, are given by

$$[\mathbf{I}]_{xx} = \int (y^2 + z^2) \rho(\mathbf{r}) d^3 r, \quad \text{and cyclic.} \tag{7.45}$$

The off-diagonal elements, called **products of inertia**, are symmetric and are given by

$$[\mathbf{I}]_{xy} = [\mathbf{I}]_{yx} = -\int xy \rho(\mathbf{r}) d^3 r, \quad \text{and cyclic.} \tag{7.46}$$

---

[1]The dyadic of inertia is more commonly called the **tensor of inertia** because it transforms like a second rank tensor under coordinate transformation.

The dyadic transforms under rotations like a second rank tensor. Since it is real and symmetric, it can be diagonalized by a rotation of coordinates, and can be represented by the diagonal matrix

$$[\mathbf{I}] = \begin{bmatrix} I_x & 0 & 0 \\ 0 & I_y & 0 \\ 0 & 0 & I_z \end{bmatrix}, \tag{7.47}$$

with the diagonal elements given by Eq. (7.45). The coordinate axes for which the dyadic of inertia is diagonal are called **principal axes**. The principal axes will generally lie along symmetry axes of any object having some symmetry.

7.8 Problems

1. Find the dipole moment of

   (a) a straight wire of length L with a linear charge density $\lambda(z) = \lambda_0 z/L$, $|z| < L/2$.

   (b) a hollow sphere of radius R with a surface charge distribution $\sigma(\theta) = (q/R^2)\cos(\theta)$.

2. Calculate the quadrupole moment for

   (a) a uniformly charged needle of length $\mathbf{L}$ and charge q.

   (b) a uniformly charged disk of radius R and charge q.

   (c) a spherical shell of radius R with surface charge distribution $(q/R^2)\cos^2\theta$.

3. Calculate the quadrupole moment for

   (a) two point charges, each of charge $+q$, a distance L apart with a third collinear point charge, -2q, at their midpoint.

   (b) a point charge, $+q$, at the center of a circular uniform line charge, -q, at radius L.

4. (a) Show, using the quadrupole dyadic $[\mathbf{Q}]$, that the potential for a symmetric quadrupole is given by

$$\phi = \frac{Q_0}{2r^3}[3(\hat{\mathbf{k}}{\cdot}\hat{\mathbf{r}})^2 - 1]. \tag{7.48}$$

   (b) Show that the electric field, given by the negative gradient of this potential, is

$$\mathbf{E} = \frac{3Q_0}{2r^4}[5(\hat{\mathbf{k}}{\cdot}\hat{\mathbf{r}})^2\hat{\mathbf{r}} - 2(\hat{\mathbf{k}}{\cdot}\hat{\mathbf{r}})\hat{\mathbf{k}} - \hat{\mathbf{r}}]. \tag{7.49}$$

5. A plywood board of mass M has dimensions $3a$ by $4a$, and a negligible thickness.

   (a) Find its principal moments of inertia.

   (b) The board rotates with angular velocity $\boldsymbol{\omega}$ about an axis along one of its diagonals. Find its angular momentum.

Chapter 8

Answered Problems

8.1 Vector Differential Operators

1. Show that $\nabla(\mathbf{r}\cdot\mathbf{p}) = \mathbf{p}$, for a constant vector $\mathbf{p}$.

    Answer: We use the definition $d\phi = d\mathbf{r}\cdot\nabla\phi$, with $\phi = \mathbf{r}\cdot\mathbf{p}$. Then

    $$d(\mathbf{r}\cdot\mathbf{p}) = d\mathbf{r}\cdot\nabla(\mathbf{r}\cdot\mathbf{p}) \quad \text{by definition.} \tag{8.1}$$
    $$d(\mathbf{r}\cdot\mathbf{p}) = d\mathbf{r}\cdot\mathbf{p} \quad \text{since only } \mathbf{r} \text{ varies.} \tag{8.2}$$

    Comparing the two equalities shows that $\nabla(\mathbf{r}\cdot\mathbf{p}) = \mathbf{p}$, since $d\mathbf{r}$ is an arbitrary displacement.

2. The potential energy between two electric dipoles, a distance $\mathbf{r} > 0$ apart is

    $$U = \frac{[3(\mathbf{p}\cdot\hat{\mathbf{r}})(\mathbf{p}'\cdot\hat{\mathbf{r}}) - \mathbf{p}\cdot\mathbf{p}']}{r^3}. \tag{8.3}$$

    Find the force between the dipoles, given by $\mathbf{F} = -\nabla U$.

    Answer: We write the expression for the energy as

    $$U = \frac{3(\mathbf{p}\cdot\mathbf{r})(\mathbf{p}'\cdot\mathbf{r})}{r^5} - \frac{\mathbf{p}\cdot\mathbf{p}'}{r^3}, \tag{8.4}$$

    because it is easier to differentiate $(\mathbf{p}\cdot\mathbf{r})$ than $(\mathbf{p}\cdot\hat{\mathbf{r}})$. Now there are four scalar functions for the gradient to act on. This gives four terms in the expression for the force between the two dipoles:

    $$\mathbf{F}_{\mathbf{p}'\mathbf{p}} = \frac{3(\mathbf{p}\cdot\hat{\mathbf{r}})\mathbf{p}' + 3(\mathbf{p}'\cdot\hat{\mathbf{r}})\mathbf{p} + 3(\mathbf{p}\cdot\mathbf{p}')\hat{\mathbf{r}} - 15(\mathbf{p}\cdot\hat{\mathbf{r}})(\mathbf{p}'\cdot\hat{\mathbf{r}})\hat{\mathbf{r}}}{r^4}. \tag{8.5}$$

We have changed $\mathbf{r}$ back into $\hat{\mathbf{r}}$, so that there will be a common denominator for the four terms.

3. Find the divergence of $\mathbf{E} = q\mathbf{r}/r^3$ for $r > 0$.

   Answer:

$$
\boldsymbol{\nabla} \cdot \left(\frac{q\mathbf{r}}{r^3}\right) = q\frac{\boldsymbol{\nabla} \cdot \mathbf{r}}{r^3} + q\mathbf{r} \cdot \boldsymbol{\nabla}\left(\frac{1}{r^3}\right)
$$

$$
= \frac{3q}{r^3} - \frac{3q(\mathbf{r} \cdot \hat{\mathbf{r}})}{r^4} = 0, \quad r > 0. \tag{8.6}
$$

   The restriction $r > 0$ is necessary because both terms in Eq. (8.6) are singular at $r = 0$.

4. Show explicitly that the curl of $\mathbf{E} = q\mathbf{r}/r^3$ vanishes.

   Answer:

$$
\boldsymbol{\nabla} \times \mathbf{E} = \boldsymbol{\nabla} \times \left[\frac{q\mathbf{r}}{r^3}\right]
$$

$$
= \frac{q}{r^3}\boldsymbol{\nabla} \times \mathbf{r} - q\mathbf{r} \times \boldsymbol{\nabla}\frac{1}{r^3}
$$

$$
= 0 + 3q\mathbf{r} \times \frac{\hat{\mathbf{r}}}{r^4} = 0. \tag{8.7}
$$

   We do not have to specify $r \neq 0$ because each term is explicitly zero for any $\mathbf{r}$. For a continuous charge distribution, the same result ($\boldsymbol{\nabla} \times \mathbf{E} = 0$) follows by bringing the $\boldsymbol{\nabla} \times$ operation inside the volume integral.

5. The potential of an electric dipole $\mathbf{p}$ is given by

$$
\phi(\mathbf{r}) = \frac{\mathbf{p} \cdot \mathbf{r}}{r^3}.
$$

   Find the electric field of the dipole (for $r > 0$), given by $\mathbf{E} = -\boldsymbol{\nabla}\phi$.

   Answer:

$$
\mathbf{E}(\mathbf{r}) = -\boldsymbol{\nabla}\left(\frac{\mathbf{p} \cdot \mathbf{r}}{r^3}\right) = -\left(\frac{1}{r^3}\right)\boldsymbol{\nabla}(\mathbf{p} \cdot \mathbf{r}) - (\mathbf{p} \cdot \mathbf{r})\boldsymbol{\nabla}\left(\frac{1}{r^3}\right)
$$

$$
= -\left(\frac{\mathbf{p}}{r^3}\right) + \left(\frac{3(\mathbf{p} \cdot \mathbf{r})\hat{\mathbf{r}}}{r^4}\right)
$$

$$
= \frac{3(\mathbf{p} \cdot \hat{\mathbf{r}})\hat{\mathbf{r}} - \mathbf{p}}{r^3}, \quad r > 0. \tag{8.8}
$$

8.2 Vector Identities

1. Derive the algebraic vector identities

  (a) $\mathbf{a}\cdot(\mathbf{b}\times\mathbf{c}) = \mathbf{b}\cdot(\mathbf{c}\times\mathbf{a}) = \mathbf{c}\cdot(\mathbf{a}\times\mathbf{b}) = (\mathbf{a}\times\mathbf{b})\cdot\mathbf{c}$.

  (b) $\mathbf{a}\times(\mathbf{b}\times\mathbf{c}) = \mathbf{b}(\mathbf{a}\cdot\mathbf{c}) - \mathbf{c}(\mathbf{a}\cdot\mathbf{b})$.

Answer:

(a) We use the fact that the cross product can be represented by the determinant

$$\mathbf{b}\times\mathbf{c} = \begin{vmatrix} \hat{\mathbf{i}} & \hat{\mathbf{j}} & \hat{\mathbf{k}} \\ b_x & b_y & b_z \\ c_x & c_y & c_z \end{vmatrix}. \qquad (8.9)$$

The triple scalar product will then be given by

$$\mathbf{a}\cdot(\mathbf{b}\times\mathbf{c}) = \begin{vmatrix} a_x & a_y & a_z \\ b_x & b_y & b_z \\ c_x & c_y & c_z \end{vmatrix}. \qquad (8.10)$$

A determinant is invariant under a cyclic permutation of its rows, which proves the symmetry properties of the triple scalar product.

(b) The triple vector product $\mathbf{a}\times(\mathbf{b}\times\mathbf{c})$ is orthogonal to the vector $\mathbf{a}$. Therefore, it must be a linear combination of the vectors $\mathbf{b}$ and $\mathbf{c}$, given as

$$\mathbf{a}\times(\mathbf{b}\times\mathbf{c}) = \beta\mathbf{b} + \gamma\mathbf{c}. \qquad (8.11)$$

Here, β and γ are scalars that must be of the form $(\mathbf{a}\cdot\mathbf{c})$ and $(\mathbf{a}\cdot\mathbf{b})$, respectively, since each term must include each of the vectors $\mathbf{a}$, $\mathbf{b}$, $\mathbf{c}$ once. This means that the triple scalar product must be a linear combination of $\mathbf{b}(\mathbf{a}\cdot\mathbf{c})$ and $\mathbf{c}(\mathbf{a}\cdot\mathbf{b})$, with only the sign of each term to be determined. We can determine the signs by choosing unit basis vectors in the triple vector product. That is, with $\mathbf{a} = \hat{\mathbf{j}}$, $\mathbf{b} = \hat{\mathbf{i}}$, $\mathbf{c} = \hat{\mathbf{j}}$,

$$\mathbf{a}\times(\mathbf{b}\times\mathbf{c}) = \hat{\mathbf{j}}\times(\hat{\mathbf{i}} \times \hat{\mathbf{j}}) = \hat{\mathbf{j}}\times\hat{\mathbf{k}} = \hat{\mathbf{i}}, \text{ and } \mathbf{b}(\mathbf{a}\cdot\mathbf{c}) = \hat{\mathbf{i}}(\hat{\mathbf{j}}\cdot\hat{\mathbf{j}}) = \hat{\mathbf{i}} \quad (8.12)$$

shows that the sign of $\mathbf{b}(\mathbf{a}\cdot\mathbf{c})$ should be positive, while

$$\hat{\mathbf{j}}\times(\hat{\mathbf{j}} \times \hat{\mathbf{i}}) = -\hat{\mathbf{j}}\times\hat{\mathbf{k}} = -\hat{\mathbf{i}} \qquad (8.13)$$

shows that the sign of $\mathbf{c}(\mathbf{a}\cdot\mathbf{b})$ should be negative. With these signs, the triple scalar product is given by

$$\mathbf{a}\times(\mathbf{b}\times\mathbf{c}) = \mathbf{b}(\mathbf{a}\cdot\mathbf{c}) - \mathbf{c}(\mathbf{a}\cdot\mathbf{b}). \qquad (8.14)$$

2. Calculate $\nabla \times (\mathbf{B} \times \mathbf{r})$ with $\mathbf{B}$ a constant vector.

   Answer:
   Use bac-cab to get

   $$\nabla \times (\mathbf{B} \times \mathbf{r}) = \mathbf{B}(\nabla \cdot \mathbf{r}) - (\mathbf{B} \cdot \nabla)\mathbf{r} = 3\mathbf{B} - \mathbf{B} = 2\mathbf{B}. \qquad (8.15)$$

3. The vector potential of a magnetic dipole $\boldsymbol{\mu}$ is given by

   $$\mathbf{A}(\mathbf{r}) \;=\; \frac{\boldsymbol{\mu} \times \mathbf{r}}{r^3}.$$

   (a) Calculate the divergence of $\mathbf{A}$.

   (b) Find the magnetic field of the dipole (for $r > 0$), given by $\mathbf{B} = \nabla \times \mathbf{A}$.

   Answer:

   (a)

   $$
   \begin{aligned}
   \nabla \cdot \mathbf{A} \;&=\; \nabla \cdot \left(\frac{\boldsymbol{\mu} \times \mathbf{r}}{r^3} \right) \\
   &=\; -\boldsymbol{\mu} \cdot \left[\nabla \times \left(\frac{\mathbf{r}}{r^3} \right) \right] \\
   &=\; 0. \qquad\qquad (8.16)
   \end{aligned}
   $$

   (b)

   $$
   \begin{aligned}
   \mathbf{B}(\mathbf{r}) \;&=\; \nabla \times \left(\frac{\boldsymbol{\mu} \times \mathbf{r}}{r^3} \right) \\
   &=\; \left(\frac{1}{r^3} \right) \nabla \times (\boldsymbol{\mu} \times \mathbf{r}) - (\boldsymbol{\mu} \times \mathbf{r}) \times \nabla \left(\frac{1}{r^3} \right) \\
   &=\; \left(\frac{1}{r^3} \right) [\boldsymbol{\mu}(\nabla \cdot \mathbf{r}) - (\boldsymbol{\mu} \cdot \nabla)\mathbf{r}] + \left(\frac{3(\boldsymbol{\mu} \times \mathbf{r}) \times \hat{\mathbf{r}}}{r^4} \right) \\
   &=\; \frac{3\boldsymbol{\mu} - \boldsymbol{\mu} + 3\hat{\mathbf{r}}(\hat{\mathbf{r}} \cdot \boldsymbol{\mu}) - 3\boldsymbol{\mu}}{r^3} \\
   &=\; \frac{3\hat{\mathbf{r}}(\hat{\mathbf{r}} \cdot \boldsymbol{\mu}) - \boldsymbol{\mu}}{r^3}, \quad r > 0. \qquad (8.17)
   \end{aligned}
   $$

4. Calculate the curl and divergence of $\mathbf{B}$, found in the previous problem.

   Answer:

   $$\nabla \times \mathbf{B}(\mathbf{r}) \;=\; \nabla \times \left[\frac{3\hat{\mathbf{r}}(\hat{\mathbf{r}} \cdot \boldsymbol{\mu}) - \boldsymbol{\mu}}{r^3} \right]$$

$$= \nabla \times \left[\frac{3\mathbf{r}(\mathbf{r}\cdot\boldsymbol{\mu})}{r^5} - \frac{\boldsymbol{\mu}}{r^3} \right]$$

$$= \frac{3(\nabla\times\mathbf{r})(\hat{\mathbf{r}}\cdot\boldsymbol{\mu}) - 3\mathbf{r}\times\nabla(\mathbf{r}\cdot\boldsymbol{\mu})}{r^5} + \boldsymbol{\mu}\times\nabla\left(\frac{1}{r^3}\right)$$

$$= \frac{-3\mathbf{r}\times\boldsymbol{\mu}}{r^5} - \frac{3\boldsymbol{\mu}\times\mathbf{r}}{r^5} = 0, \quad r > 0. \tag{8.18}$$

$$\nabla\cdot\mathbf{B}(\mathbf{r}) = \nabla\cdot\left[\frac{3\hat{\mathbf{r}}(\hat{\mathbf{r}}\cdot\boldsymbol{\mu}) - \boldsymbol{\mu}}{r^3} \right]$$

$$= \nabla\cdot\left[\frac{3\mathbf{r}(\mathbf{r}\cdot\boldsymbol{\mu})}{r^5} - \frac{\boldsymbol{\mu}}{r^3} \right]$$

$$= \frac{9(\mathbf{r}\cdot\boldsymbol{\mu}) + 3(\mathbf{r}\cdot\boldsymbol{\mu}) - 15(\mathbf{r}\cdot\boldsymbol{\mu}) + 3(\mathbf{r}\cdot\boldsymbol{\mu})}{r^5}$$

$$= 0, \quad r > 0. \tag{8.19}$$

5. Find the curl and divergence (for $r \neq 0$) of each of the vector fields:

(a) $\mathbf{F} = (\mathbf{r}\times\mathbf{p})(\mathbf{r}\cdot\mathbf{p})$,

(b) $\mathbf{G} = (\mathbf{r}\cdot\mathbf{p})^2\mathbf{r}$.

The vector $\mathbf{p}$ is a constant vector.

Answer:

(a) $\mathbf{F} = (\mathbf{r}\times\mathbf{p})(\mathbf{r}\cdot\mathbf{p})$, with $\mathbf{p}$ a constant vector.

$$\begin{aligned}
\nabla\times\mathbf{F} &= \nabla\times[(\mathbf{r}\times\mathbf{p})(\mathbf{r}\cdot\mathbf{p})] \\
&= (\mathbf{r}\cdot\mathbf{p})[\nabla\times(\mathbf{r}\times\mathbf{p})] - (\mathbf{r}\times\mathbf{p})\times[\nabla(\mathbf{r}\cdot\mathbf{p})] \\
&= (\mathbf{r}\cdot\mathbf{p})[(\mathbf{p}\cdot\nabla)\mathbf{r} - \mathbf{p}(\nabla\cdot\mathbf{r})] - (\mathbf{r}\times\mathbf{p})\times\mathbf{p} \\
&= (\mathbf{r}\cdot\mathbf{p})(\mathbf{p} - 3\mathbf{p}) - \mathbf{p}(\mathbf{r}\cdot\mathbf{p}) + p^2\mathbf{r} \\
&= p^2\mathbf{r} - 3\mathbf{p}(\mathbf{r}\cdot\mathbf{p}). \tag{8.20}
\end{aligned}$$

$$\begin{aligned}
\nabla\cdot\mathbf{F} &= \nabla\cdot[(\mathbf{r}\times\mathbf{p})(\mathbf{r}\cdot\mathbf{p})] \\
&= (\mathbf{r}\cdot\mathbf{p})[\nabla\cdot(\mathbf{r}\times\mathbf{p})] + (\mathbf{r}\times\mathbf{p})\cdot[\nabla(\mathbf{r}\cdot\mathbf{p})] \\
&= (\mathbf{r}\cdot\mathbf{p})[\mathbf{p}\cdot(\nabla\times\mathbf{r})] + (\mathbf{r}\times\mathbf{p})\cdot\mathbf{p} \\
&= 0 + \mathbf{r}\cdot(\mathbf{p}\times\mathbf{p}) = 0. \tag{8.21}
\end{aligned}$$

(b) $\mathbf{G} = (\mathbf{r \cdot p})^2 \mathbf{r}$, with $\mathbf{p}$ a constant vector.

$$
\begin{aligned}
\boldsymbol{\nabla} \times \mathbf{G} &= \boldsymbol{\nabla} \times [(\mathbf{r \cdot p})^2 \mathbf{r}] \\
&= (\mathbf{r \cdot p})^2 (\boldsymbol{\nabla} \times \mathbf{r}) - \mathbf{r} \times [\boldsymbol{\nabla} (\mathbf{r \cdot p})^2] \\
&= 0 - \mathbf{r} \times [2(\mathbf{r \cdot p})\mathbf{p}] \\
&= -2(\mathbf{r \cdot p})(\mathbf{r} \times \mathbf{p}). \qquad\qquad (8.22)
\end{aligned}
$$

$$
\begin{aligned}
\boldsymbol{\nabla} \cdot \mathbf{G} &= \boldsymbol{\nabla} \cdot [(\mathbf{r \cdot p})^2 \mathbf{r}] \\
&= (\mathbf{r \cdot p})^2 (\boldsymbol{\nabla} \cdot \mathbf{r}) + \mathbf{r} \cdot [\boldsymbol{\nabla} (\mathbf{r \cdot p})^2] \\
&= 3(\mathbf{r \cdot p})^2 + \mathbf{r} \cdot [2(\mathbf{r \cdot p})\mathbf{p}] \\
&= 5(\mathbf{r \cdot p})^2. \qquad\qquad\qquad\quad (8.23)
\end{aligned}
$$

6. Show that the operator $\mathbf{L} = -i\mathbf{r} \times \boldsymbol{\nabla}$ satisfies $\mathbf{L} \times \mathbf{L}\phi = i\mathbf{L}\phi$. [Hint: Let $\mathbf{E} = -\boldsymbol{\nabla}\phi$. Then expand $(\mathbf{r} \times \boldsymbol{\nabla}) \times (\mathbf{r} \times \mathbf{E})$ in bac-cab.]

Answer:

We want to show that $\mathbf{L} \times \mathbf{L}\phi = i\mathbf{L}\phi$, where $\mathbf{L} = -i\mathbf{r} \times \boldsymbol{\nabla}$. This corresponds to

$$
-(\mathbf{r} \times \boldsymbol{\nabla}) \times (\mathbf{r} \times \boldsymbol{\nabla})\phi = \mathbf{r} \times \boldsymbol{\nabla}\phi. \qquad (8.24)
$$

We introduce $\mathbf{E} = -\boldsymbol{\nabla}\phi$, and then will show that

$$
(\mathbf{r} \times \boldsymbol{\nabla}) \times (\mathbf{r} \times \mathbf{E}) = -\mathbf{r} \times \mathbf{E}. \qquad (8.25)
$$

We use

$$
\mathbf{a} \times (\mathbf{b} \times \mathbf{c}) = \mathbf{b}(\mathbf{a \cdot c}) - \mathbf{c}(\mathbf{a \cdot b}), \qquad (8.26)
$$

with

$$
\mathbf{r} \times \boldsymbol{\nabla} = \mathbf{a}, \quad \mathbf{r} = \mathbf{b}, \quad \mathbf{E} = \mathbf{c}. \qquad (8.27)
$$

We take the vector derivative twice, first holding $\mathbf{E}$ constant and then holding $\mathbf{r}$ constant. For $\mathbf{E}$ constant,

$$
\begin{aligned}
(\mathbf{r} \times \boldsymbol{\nabla}) \times (\mathbf{r} \times \mathbf{E}) &= [\mathbf{E} \cdot (\mathbf{r} \times \boldsymbol{\nabla})]\mathbf{r} - \mathbf{E}[(\mathbf{r} \times \boldsymbol{\nabla}) \cdot \mathbf{r}] \\
&= [(\mathbf{E} \times \mathbf{r}) \cdot \boldsymbol{\nabla})]\mathbf{r} - \mathbf{E}[\mathbf{r} \cdot (\boldsymbol{\nabla} \times \mathbf{r})] \\
&= \mathbf{E} \times \mathbf{r}. \qquad\qquad\qquad (8.28)
\end{aligned}
$$

For $\mathbf{r}$ constant,

$$
\begin{aligned}
(\mathbf{r} \times \boldsymbol{\nabla}) \times (\mathbf{r} \times \mathbf{E}) &= \mathbf{r}[(\mathbf{r} \times \boldsymbol{\nabla}) \cdot \mathbf{E}] - [\mathbf{r} \cdot (\mathbf{r} \times \boldsymbol{\nabla})]\mathbf{E} \\
&= \mathbf{r}[\mathbf{r} \cdot (\boldsymbol{\nabla} \times \mathbf{E})] - [(\mathbf{r} \times \mathbf{r}) \cdot \boldsymbol{\nabla}]\mathbf{E} \\
&= 0. \qquad\qquad\qquad\qquad (8.29)
\end{aligned}
$$

Adding the results of Eqs. (8.28) and (8.29) gives

$$(\mathbf{r}\times\boldsymbol{\nabla})\times(\mathbf{r}\times\mathbf{E}) = -\mathbf{r}\times\mathbf{E}. \qquad (8.30)$$

Then, letting $\mathbf{E} = -\boldsymbol{\nabla}\phi$,

$$-(\mathbf{r}\times\boldsymbol{\nabla})\times(\mathbf{r}\times\boldsymbol{\nabla})\phi = \mathbf{r}\times\boldsymbol{\nabla}\phi, \qquad (8.31)$$

which confirms that $\mathbf{L}\times\mathbf{L}\phi = i\mathbf{L}\phi$.

8.3 Integral Theorems

1. Demonstrate the divergence theorem

$$\int_V d^3r \boldsymbol{\nabla}\cdot\mathbf{F} \;=\; \oint_S d\mathbf{A}\cdot\mathbf{F}$$

using the function $\mathbf{F} = \mathbf{r}(\hat{\mathbf{p}}\cdot\mathbf{r})^2$ in a sphere of radius R.

Answer:

$$
\begin{aligned}
\int d^3r \boldsymbol{\nabla}\cdot\mathbf{F} &= \int d^3r \boldsymbol{\nabla}\cdot\left[\mathbf{r}(\hat{\mathbf{p}}\cdot\mathbf{r})^2\right] \\
&= \int d^3r \left\{\mathbf{r}\cdot\boldsymbol{\nabla}\left[(\mathbf{r}\cdot\hat{\mathbf{p}})^2\right] + (\mathbf{r}\cdot\hat{\mathbf{p}})^2(\boldsymbol{\nabla}\cdot\mathbf{r})\right\} \\
&= \int d^3r \left[2(\mathbf{r}\cdot\hat{\mathbf{p}})^2 + 3(\mathbf{r}\cdot\hat{\mathbf{p}})^2\right] \\
&= 5\int_0^R r^2 dr \oint d\Omega(\mathbf{r}\cdot\hat{\mathbf{p}})^2 \\
&= \frac{20\pi}{3}\int_0^R r^4 dr = \frac{4\pi R^5}{3}. \qquad (8.32)
\end{aligned}
$$

$$
\begin{aligned}
\oint_S d\mathbf{A}\cdot\mathbf{F} &= \oint_S d\mathbf{A}\cdot\left[\mathbf{r}(\hat{\mathbf{p}}\cdot\mathbf{r})^2\right] \\
&= \oint R^5 d\Omega_{\hat{\mathbf{r}}}(\hat{\mathbf{p}}\cdot\hat{\mathbf{r}})^2 = \frac{4\pi R^5}{3}. \qquad (8.33)
\end{aligned}
$$

Comparing Eqs. (8.32) and (8.33), we see that the divergence theorem is satisfied. In the above derivations, we used the result from Eq. (4.27) that

$$\langle(\hat{\mathbf{p}}\cdot\hat{\mathbf{r}})(\hat{\mathbf{r}}\cdot\hat{\mathbf{p}}')\rangle = \frac{1}{3}(\hat{\mathbf{p}}\cdot\hat{\mathbf{p}}') \qquad (8.34)$$

to do the angular integrals.

2. Demonstrate the gradient theorem

$$\int_V d^3r\nabla\phi \;=\; \oint_S d\mathbf{A}\phi$$

using the function $\phi = r(\mathbf{r}\cdot\hat{\mathbf{p}})$ in a sphere of radius R.

Answer:

$$
\begin{aligned}
\int d^3r\nabla\phi &= \int d^3r\nabla\left[r(\mathbf{r}\cdot\hat{\mathbf{p}})\right] \\
&= \int d^3r\left[r\nabla(\mathbf{r}\cdot\hat{\mathbf{p}}) + (\mathbf{r}\cdot\hat{\mathbf{p}})(\nabla r)\right] \\
&= \int d^3r\left[r\hat{\mathbf{p}} + \hat{\mathbf{r}}(\mathbf{r}\cdot\hat{\mathbf{p}})\right] \\
&= \int_0^R r^2 dr\oint d\Omega\left[r\hat{\mathbf{p}} + r\hat{\mathbf{p}}(\hat{\mathbf{p}}\cdot\hat{\mathbf{r}})((\hat{\mathbf{r}}\cdot\hat{\mathbf{p}})\right] \\
&= \frac{R^4}{4}\left(4\pi\hat{\mathbf{p}} + \frac{4\pi\hat{\mathbf{p}}}{3}\right) \\
&= \frac{4\pi R^4\hat{\mathbf{p}}}{3}.
\end{aligned}
\tag{8.35}
$$

$$
\begin{aligned}
\oint_S d\mathbf{A}\phi &= \oint_S d\mathbf{A}\left[r(\hat{\mathbf{p}}\cdot\mathbf{r})\right] \\
&= \oint R^4 d\Omega\hat{\mathbf{p}}(\hat{\mathbf{p}}\cdot\hat{\mathbf{r}})(\hat{\mathbf{p}}\cdot\hat{\mathbf{r}}) \\
&= \frac{4\pi R^4\hat{\mathbf{p}}}{3}.
\end{aligned}
\tag{8.36}
$$

Comparing Eqs. (8.35) and (8.36), we see that the gradient theorem is satisfied. We used the steps following Eq. (4.21) to do the angular integration.

3. Demonstrate the curl theorem

$$\int_V d^3r\nabla\times\mathbf{E} \;=\; \oint_S d\mathbf{A}\times\mathbf{E}$$

using the function $\mathbf{F} = \hat{\mathbf{q}}\times(\hat{\mathbf{p}}\times\mathbf{r})$ in a sphere of radius R.

Answer:

$$
\begin{aligned}
\int d^3r\nabla\times\mathbf{F} &= \int d^3r\nabla\times\left[\hat{\mathbf{q}}\times(\hat{\mathbf{p}}\times\mathbf{r})\right] \\
&= \int d^3r\nabla\times\left[\hat{\mathbf{p}}(\hat{\mathbf{q}}\cdot\mathbf{r}) - \mathbf{r}(\hat{\mathbf{q}}\cdot\hat{\mathbf{p}})\right] \\
&= -\int d^3r\hat{\mathbf{p}}\times\nabla(\hat{\mathbf{q}}\cdot\mathbf{r}) \\
&= \frac{4\pi R^3}{3}(\hat{\mathbf{q}}\times\hat{\mathbf{p}}).
\end{aligned}
\tag{8.37}
$$

$$\oint_S d\mathbf{A} \times \mathbf{F} = \oint_S d\mathbf{A} \times [\hat{\mathbf{q}} \times (\hat{\mathbf{p}} \times \mathbf{r})]$$

$$= \oint R^3 d\Omega \hat{\mathbf{r}} \times [\hat{\mathbf{p}}(\hat{\mathbf{q}} \cdot \hat{\mathbf{r}}) - \hat{\mathbf{r}}(\hat{\mathbf{q}} \cdot \hat{\mathbf{p}})]$$

$$= -\hat{\mathbf{p}} \times \oint R^3 d\Omega \hat{\mathbf{r}} (\hat{\mathbf{q}} \cdot \mathbf{r})$$

$$= \frac{4\pi R^3}{3} (\hat{\mathbf{q}} \times \hat{\mathbf{p}}). \qquad (8.38)$$

Comparing Eqs. (8.37) and (8.38), we see that the curl theorem is satisfied.

4. (a) Use Gauss's law to find the electric field inside and outside a uniformly charged hollow sphere of charge Q and radius R.

 (b) Integrate $\mathbf{E}$ to find the potential inside and outside the hollow sphere.

Answer:

 (a) By symmetry, the electric field of the uniformly charged hollow sphere is in the radial direction, and angle independent. We consider a Gaussian sphere of radius r, concentric with the hollow sphere. Then, for $r \leq R$, Gauss's law becomes

$$\oint_S \mathbf{E} \cdot d\mathbf{A} = 4\pi Q$$

$$\pi r^2 E_r = 0$$

$$\mathbf{E} = \mathbf{0}. \qquad (8.39)$$

 For $r \geq R$, Gauss's law is

$$\oint_S \mathbf{E} \cdot d\mathbf{A} = 4\pi Q$$

$$\pi r^2 E_r = 4\pi Q$$

$$\mathbf{E} = \frac{Q\hat{\mathbf{r}}}{r^2}. \qquad (8.40)$$

 (b) The electric field for $r \geq R$ is the same as that for a point charge, so the potential is the same as the potential of a point charge:

$$\phi(r) = \frac{Q}{r}. \qquad (8.41)$$

 For $r \leq R$, $\mathbf{E} = \mathbf{0}$, so the interior of a uniformly charged shell is an equipotential with

$$\phi(r) = \frac{Q}{R}. \qquad (8.42)$$

5. (a) Use Gauss's law to find the electric field inside and outside a uniformly charged solid sphere of charge Q and radius R.

   (b) Integrate $\mathbf{E}$ to find the potential inside and outside the solid sphere.

Answer:

(a) The charge density inside the uniformly charged solid sphere is

$$\rho = \frac{3Q}{4\pi R^3}. \tag{8.43}$$

By symmetry, the electric field is in the radial direction, and angle independent. We consider a Gaussian sphere of radius r, concentric with the uniformly charged sphere. Then, for $r \leq R$, Gauss's law becomes

$$\oint_S \mathbf{E} \cdot d\mathbf{A} = 4\pi Q$$
$$\pi r^2 E_r = \left[\frac{3Q}{4\pi R^3}\right]\left[\frac{4\pi r^3}{3}\right]$$
$$\mathbf{E} = \frac{Q\mathbf{r}}{R^3}. \tag{8.44}$$

For $r \geq R$, Gauss's law is

$$\oint_S \mathbf{E} \cdot d\mathbf{A} = 4\pi Q$$
$$\pi r^2 E_r = 4\pi Q$$
$$\mathbf{E} = \frac{Q\hat{\mathbf{r}}}{r^2}. \tag{8.45}$$

(b) The electric field for $r \geq R$ is the same as that for a point charge, so the potential is the same as the potential of a point charge:

$$\phi(r) = \frac{Q}{r}, \quad r \geq R. \tag{8.46}$$

For $r \leq R$,

$$\phi(r) = \phi(R) - \int_R^r \mathbf{E} \cdot d\mathbf{r} = \frac{Q}{R} - \frac{Q}{R^3}\int_R^r r\,dr$$
$$= \frac{Q}{R^3}\left[R^2 - \frac{1}{2}(r^2 - R^2)\right]$$
$$= \frac{Q}{2R^3}(3R^2 - r^2), \quad r \leq R. \tag{8.47}$$

6. (a) Use Gauss's law to find the electric field inside and outside a long straight wire of radius R with uniform charge density ρ.

   (b) Integrate $\mathbf{E}$ to find the potential inside and outside the wire with the boundary condition $\phi(R) = 0$.

Answer:

(a) The long straight wire has radius R and uniform charge density ρ. By symmetry, the electric field is perpendicular to the axis of the wire, and angle independent. We consider a Gaussian cylinder of length L and radius r, concentric with the wire. Then, for $r \leq R$, Gauss's law becomes

$$\oint_S \mathbf{E} \cdot d\mathbf{A} = 4\pi Q$$
$$2\pi r L E = 4\pi^2 r^2 L \rho$$
$$E = 2\pi \rho r, \quad r \leq R. \tag{8.48}$$

For $r \geq R$, Gauss's law is

$$\oint_S \mathbf{E} \cdot d\mathbf{A} = 4\pi Q$$
$$2\pi r L E = 4\pi^2 R^2 L \rho$$
$$E = \frac{2\pi \rho R^2}{r}, \quad r \geq R. \tag{8.49}$$

(b) We have to pick some radius for which $\phi = 0$. We choose $\phi(R) = 0$, so that the surface of the wire is at zero potential. Then, for $r \leq R$,

$$\phi(r) = -\int_R^r \mathbf{E} \cdot d\mathbf{r} = -\int_R^r 2\pi \rho r dr = \pi \rho (R^2 - r^2). \tag{8.50}$$

For $r \geq R$,

$$\phi(r) = -\int_R^r \mathbf{E} \cdot d\mathbf{r} = -\int_R^r \frac{2\pi \rho R^2 dr}{r} = 2\pi \rho R^2 \ln(R/r). \tag{8.51}$$

7. Apply Gauss's law to an infinitesimal volume, and use the divergence theorem to derive the relation

$$\nabla \cdot \mathbf{E} = 4\pi \rho.$$

Answer:

Gauss's law is

$$\oint_S d\mathbf{A}\cdot\mathbf{E} = 4\pi Q_{enclosed} = 4\pi \int \rho d^3r. \tag{8.52}$$

Applying the divergence theorem to Gauss's law, we get

$$\int \boldsymbol{\nabla}\cdot\mathbf{E} d^3r = 4\pi \int \rho d^3r. \tag{8.53}$$

For an infintesimal volume, this becomes

$$\boldsymbol{\nabla}\cdot\mathbf{E} = 4\pi\rho. \tag{8.54}$$

8.4   Dirac Delta Function

1. For the potential $\phi(r) = qe^{-\mu r}/r$,

   (a) find the electric field.

   (b) find the charge distribution that produces this potential.

   (c) show that Gauss's law is satisfied by your answers. (Use a spherical Gaussian surface and be careful about the origin.)

   Answer:

   (a) For $\phi = qe^{-\mu r}/r$, the electric field is

$$\begin{aligned}
\mathbf{E} &= -\boldsymbol{\nabla}(qe^{-\mu r}/r) = -qe^{-\mu r}\boldsymbol{\nabla}\left(\frac{1}{r}\right) - \frac{q}{r}\boldsymbol{\nabla}(e^{-\mu r}) \\
&= -qe^{-\mu r}\left(-\frac{\hat{\mathbf{r}}}{r^2} - \frac{\mu\hat{\mathbf{r}}}{r}\right) = \frac{q\hat{\mathbf{r}}e^{-\mu r}}{r^2}(1 + \mu r).
\end{aligned} \tag{8.55}$$

   (b) The charge density is given by

$$\rho = \frac{1}{4\pi}\boldsymbol{\nabla}\cdot\mathbf{E} = \frac{q}{4\pi}\boldsymbol{\nabla}\cdot\left[e^{-\mu r}\left(\frac{\mathbf{r}}{r^3} + \frac{\mu\mathbf{r}}{r^2}\right)\right], \tag{8.56}$$

   where we have written $\mathbf{E}$ in a more convenient form for taking its divergence. We differentiate each term in turn, leading to

$$\begin{aligned}
\rho &= \frac{q}{4\pi}\left\{e^{-\mu r}\left[\boldsymbol{\nabla}\cdot\left(\frac{\mathbf{r}}{r^3}\right) + \boldsymbol{\nabla}\cdot\left(\frac{\mu\mathbf{r}}{r^2}\right)\right] + \left[\frac{\mathbf{r}}{r^3} + \frac{\mu\mathbf{r}}{r^2}\right]\cdot\boldsymbol{\nabla}(e^{-\mu r})\right\} \\
&= \frac{qe^{-\mu r}}{4\pi}\left[4\pi\delta(\mathbf{r}) + \frac{3\mu}{r^2} - \frac{2\mu}{r^2} - \frac{\mu}{r^2} - \frac{\mu^2}{r}\right] \\
&= q\delta(\mathbf{r}) - \frac{\mu^2 qe^{-\mu r}}{4\pi r}.
\end{aligned} \tag{8.57}$$

Note that above we isolated the term $\nabla \cdot (\mathbf{r}/r^3)$ because we knew that it equals $4\pi\delta(\mathbf{r})$. This correctly accounted for the singular behavior at the origin. The term $q\delta(\mathbf{r})$ corresponds to a point charge q at the origin. The other part of Eq. (8.57) represents a negative charge distribution surrounding the point charge.

(c) Gauss's law is

$$\oint_S \mathbf{E}\cdot\mathbf{dA} = 4\pi Q. \tag{8.58}$$

We choose a sphere of radius R as our Gaussian surface. The integral of $\mathbf{E}\cdot\mathbf{dA}$ over the surface of the sphere is

$$\oint_S \mathbf{E}\cdot\mathbf{dA} = 4\pi R^2 E_r(R) = 4\pi q e^{-\mu R}\left(1 + \mu R\right), \tag{8.59}$$

where we have taken E_r from Eq. (8.55). The charge within the Gaussian sphere is given [using Eq. (8.57)] by

$$\begin{aligned}
Q &= \int_{r\le R} \rho d^3 r = q - 4\pi \int_0^R \frac{\mu^2 q e^{-\mu r} r^2 dr}{4\pi r} \\
&= q - q\mu^2 \int_0^R r e^{-\mu r} dr \\
&= q + q\mu^2 \left[\frac{r e^{-\mu r}}{\mu}\right]_0^R - q\mu \int_0^R e^{-\mu r} dr \\
&= q + q\mu R e^{-\mu R} - q + q e^{-\mu R} \\
&= q e^{-\mu R}\left(1 + \mu R\right), \tag{8.60}
\end{aligned}$$

which agrees with Eq. (8.59), and Gauss's law is satisfied. (The integral above was done using integration by parts.)

Note the importance of the proper treatment of the delta function at the origin.

2. Show that in the limit $a \to 0$, the function

$$f(r) = \frac{e^{-r^2/a^2}}{(\sqrt{\pi}a)^3}$$

represents the three-dimensional delta function, $\delta(\mathbf{r})$.

Answer:

$$\begin{aligned}
\text{For } r \ne 0, \quad \lim_{a\to 0} f(\mathbf{r}) &= \lim_{a\to 0}\left(\frac{e^{-r^2/a^2}}{(\sqrt{\pi}a)^3}\right) \\
&= \lim_{a\to 0}\left(\frac{a^{-3}}{(\sqrt{\pi})^3 e^{r^2/a^2}}\right) = 0, \tag{8.61}
\end{aligned}$$

using L'Hospital's rule for the limit $\frac{\infty}{\infty}$. This satisfies the delta function condition that $\int_V d^3r \delta(\mathbf{r}) = 0$ for $r \neq 0$ inside V.

Next we calculate the integral of $f(r)$ to show that it equals 1.

$$
\begin{aligned}
\int f(r)d^3r &= \int \frac{e^{-r^2/a^2} d^3r}{(\sqrt{\pi}a)^3} = \frac{4}{\sqrt{\pi}a^3} \int_0^\infty r^2 e^{-r^2/a^2} dr \\
&= \frac{2}{\sqrt{\pi}a^3} \left\{ -\left[a^2 r e^{-r^2/a^2}\right]_0^\infty + \int_0^\infty a^2 e^{-r^2/a^2} dr \right\} \\
&= \frac{2}{\sqrt{\pi}a^3} \left(\frac{\sqrt{\pi}a^3}{2} \right) = 1.
\end{aligned}
\tag{8.62}
$$

We integrated by parts and used $\int_0^\infty e^{-x^2} dx = \sqrt{\pi}/2$. Although the integral was over all space, the integral for the limit $a \to 0$ would also equal 1 over any smaller region including the origin because $\lim_{a\to 0} f(r)$ vanishes for any $r \neq 0$.

3. Show that in the limit $a \to 0$, the function

$$
f(x) = \frac{e^{-x^2/a^2}}{a\sqrt{\pi}}
$$

represents the one-dimensional delta function, $\delta(x)$.

Answer:

$$
\begin{aligned}
\text{For } x \neq 0, \quad \lim_{a\to 0} f(x) &= \lim_{a\to 0} \left(\frac{e^{-x^2/a^2}}{a\sqrt{\pi}} \right) \\
&= \lim_{a\to 0} \left(\frac{a^{-1}}{e^{x^2/a^2}\sqrt{\pi}} \right) = 0,
\end{aligned}
\tag{8.63}
$$

which we can confirm by L'Hospital's rule for the limit $\frac{\infty}{\infty}$. This satisfies the delta function condition that $\delta(x) = 0$ for $x \neq 0$. Next we calculate the integral of $f(x)$ over all x to show that it equals 1.

$$
\int_{-\infty}^{+\infty} dx f(x) = \frac{1}{a\sqrt{\pi}} \int_{-\infty}^{+\infty} e^{-x^2/a^2} dx = 1.
\tag{8.64}
$$

Although this integral was over all x, the integral for the limit $a \to 0$ would also equal 1 over any smaller region including the origin because $\lim_{a\to 0} f(x)$ vanishes for any $x \neq 0$.

4. Show that

$$
f(x) = \lim_{\epsilon \to 0} \left[\text{Im} \left(\frac{1}{x - i\epsilon} \right) \right] = \pi \delta(x).
\tag{8.65}
$$

Answer:

$$
\begin{aligned}
f(x) &= \lim_{\epsilon \to 0}\left[\operatorname{Im}\left(\frac{1}{x - i\epsilon}\right)\right] \\
&= \lim_{\epsilon \to 0}\left[\operatorname{Im}\left(\frac{x + i\epsilon}{x^2 + \epsilon^2}\right)\right] \\
&= \lim_{\epsilon \to 0}\left(\frac{\epsilon}{x^2 + \epsilon^2}\right) \\
&= 0, \quad x \neq 0.
\end{aligned}
\tag{8.66}
$$

Next we calculate the $\int_{-\infty}^{+\infty} f(x)dx$ to show that it equals π.

$$
\begin{aligned}
\int_{-\infty}^{+\infty} f(x)dx &= \lim_{\epsilon \to 0}\int_{-\infty}^{+\infty}\frac{\epsilon dx}{x^2 + \epsilon^2} \\
&= \lim_{\epsilon \to 0}\int_{-\pi/2}^{+\pi/2} d\theta = \pi,
\end{aligned}
\tag{8.67}
$$

where we made the substitution $x = \epsilon \tan\theta$ in the integral. Although this integral was from $-\infty$ to $+\infty$, the integral would also equal 1 over any region including the origin because $\left(\frac{\epsilon}{x^2+\epsilon^2}\right)$ vanishes for any $x \neq 0$ in the limit $\epsilon \to 0$.

5. Show that

$$
\delta[f(x)] = \sum_i \frac{\delta(x - x_i)}{\left|\frac{df}{dx}\right|_{x=x_i}},
$$

where the x_i are the zeros of $f(x)$.

Answer:

In the integral $\int_a^b g(x)\delta[f(x)]dx$, we change the integration variable from x to $f(x)$, resulting in

$$
\begin{aligned}
\int_a^b g(x)\delta[f(x)] &= \int_{f(a)}^{f(b)} \frac{g(x)\delta[f(x)]d[f(x)]}{\left|\frac{df}{dx}\right|} \\
&= \sum_i \frac{g(x_i)}{\left|\frac{df}{dx}\right|_{x=x_i}},
\end{aligned}
\tag{8.68}
$$

where the x_i are the zeros of $f(x)$ in the range $a < x < b$. This establishes that

$$
\delta[f(x)] = \sum_i \frac{\delta(x - x_i)}{\left|\frac{df}{dx}\right|_{x=x_i}}, \quad f(x_i) = 0.
\tag{8.69}
$$

8.5   Green's Functions

1. (a) Show that the function

$$G(\boldsymbol{\rho}, z; \boldsymbol{\rho}', z') \; = \; \frac{1}{|\boldsymbol{\rho} - \boldsymbol{\rho}' + (z - z')\hat{\mathbf{k}}|} - \frac{1}{|\boldsymbol{\rho} - \boldsymbol{\rho}' - (z + z')\hat{\mathbf{k}}|}$$

defined in the half plane $z, z' \geq 0$, satisfies the two criteria for a Dirichlet Green's function. The vectors $\boldsymbol{\rho}$ and $\boldsymbol{\rho}'$ are two-dimensional vectors in the $z, z' = 0$ plane.

   (b) Find the surface Green's function for this Green's function.

   (c) Use the surface Green's function to find the potential along the positive z axis ($\rho = 0$) for the boundary condition

$$\begin{aligned}
\phi(\boldsymbol{\rho}', 0) &= V. \quad \rho' < a \\
&= 0, \quad \rho' > a.
\end{aligned} \tag{8.70}$$

Answer:

   (a) For $z' = 0$,

$$\begin{aligned}
G(\boldsymbol{\rho}, z; \boldsymbol{\rho}', 0) &= \frac{1}{|\boldsymbol{\rho} - \boldsymbol{\rho}' + z\hat{\mathbf{k}}|} - \frac{1}{|\boldsymbol{\rho} - \boldsymbol{\rho}' - z\hat{\mathbf{k}}|} \\
&= \frac{1}{[|\boldsymbol{\rho} - \boldsymbol{\rho}'|^2 + z^2]^{\frac{1}{2}}} - \frac{1}{[|\boldsymbol{\rho} - \boldsymbol{\rho}'|^2 + z^2]^{\frac{1}{2}}} = 0. \,(8.71)
\end{aligned}$$

This result satisfies the homogeneous Dirichlet boundary condition for a Green's function.

   The Laplacian of the first term of $G(\boldsymbol{\rho}, z; \boldsymbol{\rho}', z')$ results in

$$\nabla'^2 \left(\frac{1}{|\mathbf{r} - \mathbf{r}'|} \right) = -4\pi\delta(\mathbf{r}' - \mathbf{r}). \tag{8.72}$$

The Laplacian of the second term vanishes because its denominator cannot become zero.

   We see that the function $G(\boldsymbol{\rho}, z; \boldsymbol{\rho}', z')$ satisfies the two criteria of a Dirichlet Green's function. It vanishes on the bounding surface, and its Laplacian is proportional to the Dirac delta function.

   (b) The surface Green's function with $\rho = 0$ is given by

$$g(\mathbf{0}, z; \boldsymbol{\rho}', 0) = \frac{1}{4\pi}\partial_{z'}\left\{ \frac{1}{[(\rho'^2 + (z' - z)^2]^{\frac{1}{2}}} - \frac{1}{[\rho'^2 + (z + z')^2]^{\frac{1}{2}}} \right\}_{z'=0}$$

$$= \frac{1}{4\pi} \left\{ \frac{(z-z')}{[(\rho'^2 + (z'-z)^2]^{\frac{3}{2}}} + \frac{(z+z')}{[\rho'^2 + (z+z')^2]^{\frac{3}{2}}} \right\}_{z'=0}$$

$$= \frac{z}{2\pi(\rho'^2 + z^2)^{\frac{3}{2}}}. \qquad (8.73)$$

We have used a positive sign in the first step above instead of the negative sign in Eq. (5.10) defining the surface Green's function because the positive z direction here is into the volume instead of being directed out of the volume.

(c) The potential on the z axis, corresponding to the boundary condition on $\phi(\boldsymbol{\rho}, 0)$ is

$$
\begin{aligned}
\phi(\mathbf{0}, z) &= \int g(\mathbf{0}, z; \boldsymbol{\rho}', 0) \phi(\boldsymbol{\rho}', 0) dA' \\
&= V \int_0^a g(\mathbf{0}, z; \boldsymbol{\rho}', 0) 2\pi \rho' d\rho' \\
&= Vz \int_0^a \frac{\rho' d\rho'}{(\rho'^2 + z^2)^{\frac{3}{2}}} \\
&= V \left[1 - \frac{z}{\sqrt{z^2 + a^2}} \right]. \qquad (8.74)
\end{aligned}
$$

2. Show that the function

$$G(\mathbf{r}, \mathbf{r}') = \frac{1}{|\mathbf{r}' - \mathbf{r}|} - \frac{(R/r)}{|\mathbf{r}' - (R/r)^2 \mathbf{r}|}, \qquad r, r' \leq R \qquad (8.75)$$

defined inside a sphere of radius R, satisfies the two criteria for a Dirichlet Green's function.

Answer:
For $\mathbf{r}'$ on the surface of the bounding sphere,

$$
\begin{aligned}
G(\mathbf{r}, R\hat{\mathbf{r}}') &= \frac{1}{|R\hat{\mathbf{r}}' - \mathbf{r}|} - \frac{(R/r)}{|R\hat{\mathbf{r}}' - (R/r)^2 \mathbf{r}|} \\
&= \frac{1}{|R\hat{\mathbf{r}}' - r\hat{\mathbf{r}}|} - \frac{1}{|r\hat{\mathbf{r}}' - R\hat{\mathbf{r}}|} \\
&= 0 \quad \text{by the law of cosines.} \qquad (8.76)
\end{aligned}
$$

This result satisfies the homogeneous Dirichlet boundary condition for a Green's function.

The Laplacian of the first term of $G(\mathbf{r}, \mathbf{r}'))$ results in

$$\nabla'^2 \left(\frac{1}{|\mathbf{r} - \mathbf{r}'|} \right) = -4\pi\delta(\mathbf{r}' - \mathbf{r}). \qquad (8.77)$$

The Laplacian of the second term vanishes because its denominator cannot become zero. We see that the function $G(\mathbf{r}, \mathbf{r}')$ satisfies the two criteria of a Dirichlet Green's function. It vanishes on the bounding surface, and its Laplacian is proportional to the Dirac delta function.

8.6   General Coordinate Systems

1. (a) Write the potential of a dipole $\mathbf{p}$ in Cartesian coordinates.

   (b) Find $\mathbf{E}$ of the dipole for $r > 0$ by taking the negative gradient in Cartesian coordinates. Show that this equals the result of writing the vector expression for $\mathbf{E}$ in Cartesian coordinates.

   (c) Calculate the curl and the divergence of $\mathbf{E}$ of the dipole in Cartesian coordinates for $r > 0$. (Each should come out zero.)

   Answer

   (a) In Cartesian coordinates, the potential of a dipole is

$$\begin{aligned} \psi(\mathbf{r}) &= \frac{\mathbf{p}\cdot\mathbf{r}}{r^3} \\ &= \frac{xp_x + yp_y + zp_z}{(x^2 + y^2 + z^2)^{\frac{3}{2}}}. \end{aligned} \qquad (8.78)$$

   (b) The Cartesian coordinates of the electric field are given by

$$\begin{aligned} E_x &= \partial_x\psi = -\partial_x \left[\frac{xp_x + yp_y + zp_z}{(x^2 + y^2 + z^2)^{\frac{3}{2}}} \right] \\ &= -\frac{p_x}{(x^2 + y^2 + z^2)^{\frac{3}{2}}} + \frac{3x(xp_x + yp_y + zp_z)}{(x^2 + y^2 + z^2)^{\frac{5}{2}}}, \text{ and cyclic. } (8.79) \end{aligned}$$

   The vector expression for the electric field of the dipole is

$$\mathbf{E} = \frac{3(\mathbf{p}\cdot\hat{\mathbf{r}})\hat{\mathbf{r}} - \mathbf{p}}{r^3}. \qquad (8.80)$$

   The Cartesian components of this are

$$E_x = \frac{3x(xp_x + yp_y + zp_z)}{(x^2 + y^2 + z^2)^{\frac{5}{2}}} - \frac{p_x}{(x^2 + y^2 + z^2)^{\frac{3}{2}}}, \text{ and cyclic. } (8.81)$$

   These Cartesian components are the same as those given by Eq. (8.79).

(c) The Cartesian coordinates of the curl of $\mathbf{E}$ are

$$(\mathbf{\nabla} \times \mathbf{E})_x \;=\; \partial_x E_y - \partial_y E_x, \quad \text{and cyclic} \tag{8.82}$$

We look first at the second term in Eq. (8.82)

$$
\begin{aligned}
-\partial_y E_x \;=\; & -\partial_y \left[\frac{3x(xp_x + yp_y + zp_z)}{(x^2 + y^2 + z^2)^{\frac{5}{2}}} - \frac{p_x}{(x^2 + y^2 + z^2)^{\frac{3}{2}}} \right] \\
\;=\; & \frac{15xy(xp_x + yp_y + zp_z)}{(x^2 + y^2 + z^2)^{\frac{7}{2}}} \\
& - \frac{3xp_y}{(x^2 + y^2 + z^2)^{\frac{5}{2}}} - \frac{3yp_x}{(x^2 + y^2 + z^2)^{\frac{5}{2}}}.
\end{aligned}
\tag{8.83}
$$

This result is symmetric in x and y, so the antisymmetric combination

$$\partial_x E_y - \partial_y E_x = 0, \tag{8.84}$$

and $\mathbf{\nabla} \times \mathbf{E} = 0$ in Cartesian coordinates.

The divergence of $\mathbf{E}$ in Cartesian coordinates is

$$
\begin{aligned}
\mathbf{\nabla} \cdot \mathbf{E} \;=\; & \partial_x E_x + \partial_y E_y + \partial_z E_z \\
\;=\; & \partial_x \left[\frac{3x(xp_x + yp_y + zp_z)}{(x^2 + y^2 + z^2)^{\frac{5}{2}}} - \frac{p_x}{(x^2 + y^2 + z^2)^{\frac{3}{2}}} \right] \quad \text{plus cyclic.} \\
\;=\; & \frac{[6xp_x + 3(yp_y + zp_z)]}{(x^2 + y^2 + z^2)^{\frac{5}{2}}} - \frac{15x^2(xp_x + yp_y + zp_z)}{(x^2 + y^2 + z^2)^{\frac{7}{2}}} \\
& + \frac{3xp_x}{(x^2 + y^2 + z^2)^{\frac{5}{2}}} \quad \text{plus cyclic.} \\
\;=\; & \frac{[9xp_x + 3(yp_y + zp_z)](x^2 + y^2 + z^2) - 15x^2(xp_x + yp_y + zp_z)}{(x^2 + y^2 + z^2)^{\frac{7}{2}}} \\
& + \frac{[9yp_y + 3(zp_z + xp_x)](x^2 + y^2 + z^2) - 15y^2(xp_x + yp_y + zp_z)}{(x^2 + y^2 + z^2)^{\frac{7}{2}}} \\
& + \frac{[9zp_z + 3(xp_x + yp_y)](x^2 + y^2 + z^2) - 15z^2(xp_x + yp_y + zp_z)}{(x^2 + y^2 + z^2)^{\frac{7}{2}}}.
\end{aligned}
\tag{8.85}
$$

All of the terms in Eq. (8.85) now have the same denominator. We give below all the terms of the numerator that involve p_x:

$$
\begin{aligned}
p_x(& 9x^3 + 3x^3 + 3x^3 - 15x^3 \\
& + 9xy^2 + 3xy^2 + 3xy^2 - 15xy^2 \\
& + 9xz^2 + 3xz^2 + 3xz^2 - 15xz^2) \\
= \; & 0.
\end{aligned}
\tag{8.86}
$$

Using the same method, the sum of terms containing p_y and those containing p_z also cancel, so the divergence of $\mathbf{E}$ is zero.

2. (a) Verify algebraically that spherical coordinates form an orthogonal coordinate system.

    (b) Calculate the metric coefficients h_i for spherical coordinates algebraically.

Answer:

(a) The orthogonality relation for the metric coefficients is

$$\sum_k \frac{\partial x_k}{\partial q_i} \frac{\partial x_k}{\partial q_j} = h_i h_j \delta_{ij} = 0 \text{ for } i \neq j. \tag{8.87}$$

There are three independent orthogonality relations. These are

1. $\dfrac{\partial x}{\partial r} \dfrac{\partial x}{\partial \theta} + \dfrac{\partial y}{\partial r} \dfrac{\partial y}{\partial \theta} + \dfrac{\partial z}{\partial r} \dfrac{\partial z}{\partial \theta}$

$$= \left[\frac{\partial(r \sin \theta \cos \phi)}{\partial r}\right] \left[\frac{\partial(r \sin \theta \cos \phi)}{\partial \theta}\right]$$

$$+ \left[\frac{\partial(r \sin \theta \sin \phi)}{\partial r}\right] \left[\frac{\partial(r \sin \theta \sin \phi)}{\partial \theta}\right]$$

$$+ \left[\frac{\partial(r \cos \theta)}{\partial r}\right] \left[\frac{\partial(r \cos \theta)}{\partial \theta}\right]$$

$$= r \sin \theta \cos \theta (\cos^2 \phi + \sin^2 \phi - 1) = 0. \tag{8.88}$$

This shows that $\hat{\mathbf{r}} \cdot \hat{\boldsymbol{\theta}} = 0$.

2. $\dfrac{\partial x}{\partial r} \dfrac{\partial x}{\partial \phi} + \dfrac{\partial y}{\partial r} \dfrac{\partial y}{\partial \phi} + \dfrac{\partial z}{\partial r} \dfrac{\partial z}{\partial \phi}$

$$= \left[\frac{\partial(r \sin \theta \cos \phi)}{\partial r}\right] \left[\frac{\partial(r \sin \theta \cos \phi)}{\partial \phi}\right]$$

$$+ \left[\frac{\partial(r \sin \theta \sin \phi)}{\partial r}\right] \left[\frac{\partial(r \sin \theta \sin \phi)}{\partial \phi}\right]$$

$$+ \left[\frac{\partial(r \cos \theta)}{\partial r}\right] \left[\frac{\partial(r \cos \theta)}{\partial \phi}\right]$$

$$= -r \sin^2 \theta \cos \phi \sin \phi + r \sin^2 \theta \sin \phi \cos \phi + 0 = 0. \tag{8.89}$$

This shows that $\hat{\mathbf{r}} \cdot \hat{\boldsymbol{\phi}} = 0$.

3. $\dfrac{\partial x}{\partial \theta} \dfrac{\partial x}{\partial \phi} + \dfrac{\partial y}{\partial \theta} \dfrac{\partial y}{\partial \phi} + \dfrac{\partial z}{\partial \theta} \dfrac{\partial z}{\partial \phi}$

$$= \left[\frac{\partial(r \sin \theta \cos \phi)}{\partial \theta} \right] \left[\frac{\partial(r \sin \theta \cos \phi)}{\partial \phi} \right]$$

$$+ \left[\frac{\partial(r \sin \theta \sin \phi)}{\partial \theta} \right] \left[\frac{\partial(r \sin \theta \sin \phi)}{\partial \phi} \right]$$

$$+ \left[\frac{\partial(r \cos \theta)}{\partial \theta} \right] \left[\frac{\partial(r \cos \theta)}{\partial \phi} \right]$$

$$= -r^2 \cos \theta \sin \theta \cos \phi \sin \phi + r^2 \cos \theta \sin \theta \sin \phi \cos \phi + 0 = 0. (8.90)$$

This shows that $\hat{\boldsymbol{\theta}} \cdot \hat{\boldsymbol{\phi}} = 0$.

(b) For $j = i$, Eq. (8.87) gives the square of the metric coefficient h_i:

$$h_i^2 = \sum_k \left(\frac{\partial x_k}{\partial q_i} \right)^2. \tag{8.91}$$

This gives

$$h_r^2 = \left(\frac{\partial x}{\partial r} \right)^2 + \left(\frac{\partial y}{\partial r} \right)^2 + \left(\frac{\partial z}{\partial r} \right)^2$$

$$= \left[\frac{\partial(r \sin \theta \cos \phi)}{\partial r} \right]^2 + \left[\frac{\partial(r \sin \theta \sin \phi)}{\partial r} \right]^2 + \left[\frac{\partial(r \cos \theta)}{\partial r} \right]^2$$

$$= \sin^2 \theta \cos^2 \phi + \sin^2 \theta \sin^2 \phi + \cos^2 \theta = 1,$$

so $h_r = 1;$ \hfill (8.92)

$$h_\theta^2 = \left(\frac{\partial x}{\partial \theta} \right)^2 + \left(\frac{\partial y}{\partial \theta} \right)^2 + \left(\frac{\partial z}{\partial \theta} \right)^2$$

$$= \left[\frac{\partial(r \sin \theta \cos \phi)}{\partial \theta} \right]^2 + \left[\frac{\partial(r \sin \theta \sin \phi)}{\partial \theta} \right]^2 + \left[\frac{\partial(r \cos \theta)}{\partial \theta} \right]^2$$

$$= r^2 \cos^2 \theta \cos^2 \phi + r^2 \cos^2 \theta \sin^2 \phi + r^2 \sin^2 \theta = r^2,$$

so $h_\theta = r;$ \hfill (8.93)

$$h_\phi^2 = \left(\frac{\partial x}{\partial \phi} \right)^2 + \left(\frac{\partial y}{\partial \phi} \right)^2 + \left(\frac{\partial z}{\partial \phi} \right)^2$$

$$= \left[\frac{\partial(r \sin \theta \cos \phi)}{\partial \phi} \right]^2 + \left[\frac{\partial(r \sin \theta \sin \phi)}{\partial \phi} \right]^2 + \left[\frac{\partial(r \cos \theta)}{\partial \phi} \right]^2$$

$$= r^2 \sin^2 \theta \sin^2 \phi + r^2 \sin^2 \theta \cos^2 \phi + 0 = r^2 \sin^2 \theta,$$

so $h_\phi = r \sin \theta.$ \hfill (8.94)

3. (a) Write the potential of a dipole **p** (pointing in the z direction) in spherical coordinates.

   (b) Find **E** of the dipole for $r > 0$ by taking the negative gradient in spherical coordinates. Show that this equals the result of writing the vector expression for **E** in spherical coordinates.

   (c) Calculate the curl and the divergence of **E** of the dipole in spherical coordinates for $r > 0$. (Each should come out zero.)

Answer:

In spherical coordinates, the potential of an electric dipole pointing in the z direction is

(a)

$$\psi = \frac{\mathbf{p}\cdot\hat{\mathbf{r}}}{r^2} = \frac{p\cos\theta}{r^2}. \tag{8.95}$$

(b) The spherical coordinates of the electric field are given by

$$E_r = -\partial_r\left(\frac{p\cos\theta}{r^2}\right) = \frac{2p\cos\theta}{r^3}, \tag{8.96}$$

$$E_\theta = -\frac{\partial_\theta(p\cos\theta)}{r^3} = \frac{p\sin\theta}{r^3}, \tag{8.97}$$

$$E_\phi = -\frac{\partial_\phi(p\cos\theta)}{r^3\sin\theta} = 0. \tag{8.98}$$

The vector expression for **E** due to an electric dipole is

$$\mathbf{E} = \frac{3(\mathbf{p}\cdot\hat{\mathbf{r}})\hat{\mathbf{r}} - \mathbf{p}}{r^3}. \tag{8.99}$$

The spherical components of this are

$$E_r = \hat{\mathbf{r}}\cdot\mathbf{E} = \frac{\hat{\mathbf{r}}\cdot[3(\mathbf{p}\cdot\hat{\mathbf{r}})\hat{\mathbf{r}} - \mathbf{p}]}{r^3} = \frac{2p\cos\theta}{r^3}, \tag{8.100}$$

$$E_\theta = \hat{\boldsymbol{\theta}}\cdot\mathbf{E} = \frac{\hat{\boldsymbol{\theta}}\cdot[3(\mathbf{p}\cdot\hat{\mathbf{r}})\hat{\mathbf{r}} - \mathbf{p}]}{r^3} = \frac{-\hat{\boldsymbol{\theta}}\cdot\mathbf{p}}{r^3} = \frac{p\sin\theta}{r^3}, \tag{8.101}$$

$$E_\phi = \hat{\boldsymbol{\phi}}\cdot\frac{[3(\mathbf{p}\cdot\hat{\mathbf{r}})\hat{\mathbf{r}} - \mathbf{p}]}{r^3} = 0. \tag{8.102}$$

These spherical components agree with those in Eqs. (8.96)-(8.98).

(c) The spherical coordinates of the curl of **E** are

$$(\boldsymbol{\nabla}\times\mathbf{E})_r = \frac{\partial_\theta(r\sin\theta\,E_\phi) - \partial_\phi(r E_\theta)}{r^2\sin\theta} = 0, \tag{8.103}$$

$$(\nabla \times \mathbf{E})_\theta = \frac{\partial_\phi E_r - \partial_r(r\sin\theta E_\phi)}{r\sin\theta} = 0, \tag{8.104}$$

$$(\nabla \times \mathbf{E})_\phi = \frac{\partial_r(rE_\theta) - \partial_\theta E_r}{r}$$

$$= \frac{\partial_r(p\sin\theta/r^2) - \partial_\theta(2p\cos\theta/r^3)}{r}$$

$$= \frac{-2p\sin\theta/r^3 + 2p\sin\theta/r^3}{r} = 0. \tag{8.105}$$

The divergence of $\mathbf{E}$ in spherical coordinates is

$$\nabla \cdot \mathbf{E} = \frac{1}{r^2\sin\theta}\left[\partial_r(r^2\sin\theta E_r) + \partial_\theta(r\sin\theta E_\theta) + \partial_\phi(rE_\phi)\right]$$

$$= \frac{1}{r^2}\partial_r\left(\frac{2p\cos\theta}{r}\right) + \frac{1}{r\sin\theta}\partial_\theta\left(\frac{p\sin^2\theta}{r^3}\right) + 0$$

$$= \frac{-2p\cos\theta}{r^4} + \frac{2p\cos\theta}{r^4} = 0. \tag{8.106}$$

4. (a) Write the potential of an electric dipole $\mathbf{p}$ (pointing in a general direction) in spherical coordinates.

(b) Find $\mathbf{E}$ of the dipole for $r > 0$ by taking the negative gradient in spherical coordinates. Show that this equals the result of writing the vector expression for $\mathbf{E}$ in spherical coordinates.

(c) Calculate the curl and the divergence of $\mathbf{E}$ of the dipole in spherical coordinates for $r > 0$. (Each should come out zero.)

Answer:

(a) In spherical coordinates, the potential of an electric dipole is

$$\psi = \frac{\mathbf{p}\cdot\hat{\mathbf{r}}}{r^2} = \frac{p_r}{r^2}. \tag{8.107}$$

(b) The spherical coordinates of the electric field, $\mathbf{E} = -\nabla\psi$, are given by

$$E_r = -\frac{\partial_r p_r}{r^2}, \tag{8.108}$$

$$E_\theta = -\frac{\partial_\theta p_r}{r^3}, \tag{8.109}$$

$$E_\phi = -\frac{\partial_\phi p_r}{r^3\sin\theta}. \tag{8.110}$$

To continue, we have to evaluate the partial derivatives $\partial_r(p_r)$, $\partial_\theta(p_r)$, and $\partial_\phi(p_r)$. We do this by evaluating the gradient of p_r in two different ways:

$$
\begin{aligned}
1. \quad \nabla p_r &= \nabla\left(\frac{\mathbf{p}\cdot\mathbf{r}}{r}\right) = \frac{\mathbf{p}}{r} - \frac{(\mathbf{p}\cdot\mathbf{r})\hat{\mathbf{r}}}{r^2} \\
&= \frac{\hat{\mathbf{r}}p_r}{r} + \frac{\hat{\boldsymbol{\theta}}p_\theta}{r} + \frac{\hat{\boldsymbol{\phi}}p_\phi}{r} - \frac{\hat{\mathbf{r}}p_r}{r} \\
&= \frac{p_\theta\hat{\boldsymbol{\theta}}}{r} + \frac{p_\phi\hat{\boldsymbol{\phi}}}{r}.
\end{aligned}
\tag{8.111}
$$

We also have the direct expansion of ∇p_r:

$$
2. \quad \nabla p_r = (\partial_r p_r)\hat{\mathbf{r}} + \frac{(\partial_\theta p_r)\hat{\boldsymbol{\theta}}}{r} + \frac{(\partial_\phi p_r)\hat{\boldsymbol{\phi}}}{r\sin\theta}.
\tag{8.112}
$$

Comparing these two equations, we see that

$$
\begin{aligned}
\partial_r p_r &= 0, & (8.113) \\
\partial_\theta p_r &= p_\theta, & (8.114) \\
\partial_\phi p_r &= p_\phi\sin\theta. & (8.115)
\end{aligned}
$$

Now we can substitute Eqs. (8.113)-(8.115) into each of Eqs. (8.108)-(8.110) to get

$$
E_r = -\frac{\partial_r p_r}{r^2} = \frac{2p_r}{r^3},
\tag{8.116}
$$

$$
E_\theta = -\frac{\partial_\theta p_r}{r^3} = -\frac{p_\theta}{r^3},
\tag{8.117}
$$

$$
E_\phi = -\frac{\partial_\phi p_r}{r^3\sin\theta} = -p_\phi/r^3.
\tag{8.118}
$$

The vector expression for $\mathbf{E}$ due to an electric dipole is

$$
\mathbf{E} = \frac{3(\mathbf{p}\cdot\hat{\mathbf{r}})\hat{\mathbf{r}} - \mathbf{p}}{r^3}.
\tag{8.119}
$$

The spherical components of this are

$$
E_r = \hat{\mathbf{r}}\cdot\mathbf{E} = \frac{\hat{\mathbf{r}}\cdot[3(\mathbf{p}\cdot\hat{\mathbf{r}})\hat{\mathbf{r}} - \mathbf{p}]}{r^3} = \frac{2p_r}{r^3},
\tag{8.120}
$$

$$
E_\theta = \hat{\boldsymbol{\theta}}\cdot\mathbf{E} = \frac{\hat{\boldsymbol{\theta}}\cdot[3(\mathbf{p}\cdot\hat{\mathbf{r}})\hat{\mathbf{r}} - \mathbf{p}]}{r^3} = \frac{-\hat{\boldsymbol{\theta}}\cdot\mathbf{p}}{r^3} = \frac{-p_\theta}{r^3},
\tag{8.121}
$$

$$
E_\phi = \hat{\boldsymbol{\phi}}\cdot\mathbf{E} = \frac{-\hat{\boldsymbol{\phi}}\cdot\mathbf{p}}{r^3} = \frac{-p_\phi}{r^3}.
\tag{8.122}
$$

These spherical components agree with those in Eqs. (8.116)-(8.118).

5. (a) Verify algebraically that cylindrical coordinates form an orthogonal coordinate system.

   (b) Calculate the metric coefficients h_i for cylindrical coordinates algebraically.

Answer:

(a) The orthogonality relation for the metric coefficients is

$$\sum_k \frac{\partial x_k}{\partial q_i}\frac{\partial x_k}{\partial q_j} = h_i h_j \delta_{ij} = 0 \text{ for } i \neq j. \tag{8.123}$$

For cylindrical coordinates, we will use the designation ρ for the radial coordinate. This is to distinguish it from the vector $\mathbf{r}$ which designates the vector distance from the origin, and is related to the radial vector $\boldsymbol{\rho}$ by $\mathbf{r} = \rho\hat{\boldsymbol{\rho}} + z\hat{\mathbf{k}}$.

There are three independent orthogonality relations. These are

1. $\dfrac{\partial x}{\partial \rho}\dfrac{\partial x}{\partial \theta} + \dfrac{\partial y}{\partial \rho}\dfrac{\partial y}{\partial \theta} + \dfrac{\partial z}{\partial \rho}\dfrac{\partial z}{\partial \theta} = \left[\dfrac{\partial(\rho\cos\theta)}{\partial\rho}\right]\left[\dfrac{\partial(\rho\cos\theta)}{\partial\theta}\right]$

$$+\left[\frac{\partial(\rho\sin\theta)}{\partial\rho}\right]\left[\frac{\partial(\rho\sin\theta)}{\partial\theta}\right] + \left[\frac{\partial z}{\partial\rho}\right]\left[\frac{\partial z}{\partial\theta}\right]$$

$$= -\rho\cos\theta\sin\theta + \rho\sin\theta\cos\theta + 0 = 0. \tag{8.124}$$

This shows that $\hat{\boldsymbol{\rho}}\cdot\hat{\boldsymbol{\theta}} = 0$.

2. $\dfrac{\partial x}{\partial \rho}\dfrac{\partial x}{\partial z} + \dfrac{\partial y}{\partial \rho}\dfrac{\partial y}{\partial z} + \dfrac{\partial z}{\partial \rho}\dfrac{\partial z}{\partial z} = \left[\dfrac{\partial(\rho\cos\theta)}{\partial\rho}\right]\left[\dfrac{\partial(\rho\cos\theta)}{\partial z}\right]$

$$+\left[\frac{\partial(\rho\sin\theta)}{\partial\rho}\right]\left[\frac{\partial(\rho\sin\theta)}{\partial z}\right] + \left[\frac{\partial z}{\partial\rho}\right]\left[\frac{\partial z}{\partial z}\right] = 0. \tag{8.125}$$

This shows that $\hat{\boldsymbol{\rho}}\cdot\hat{\mathbf{k}} = 0$.

3. $\dfrac{\partial x}{\partial \theta}\dfrac{\partial x}{\partial z} + \dfrac{\partial y}{\partial \theta}\dfrac{\partial y}{\partial z} + \dfrac{\partial z}{\partial \theta}\dfrac{\partial z}{\partial z} = \left[\dfrac{\partial(\rho\cos\theta)}{\partial\theta}\right]\left[\dfrac{\partial(\rho\cos\theta)}{\partial z}\right]$

$$+\left[\frac{\partial(\rho\sin\theta)}{\partial\theta}\right]\left[\frac{\partial(\rho\sin\theta)}{\partial z}\right] + \left[\frac{\partial z}{\partial\theta}\right]\left[\frac{\partial z}{\partial z}\right] = 0. \tag{8.126}$$

This shows that $\hat{\boldsymbol{\theta}}\cdot\hat{\mathbf{k}} = 0$.

(b) For $j = i$, Eq. (8.87) gives the square of the metric coefficient h_i:

$$h_i^2 = \sum_k \left(\frac{\partial x_k}{\partial q_i}\right)^2. \tag{8.127}$$

This gives

$$
\begin{aligned}
h_\rho^2 &= \left(\frac{\partial x}{\partial \rho}\right)^2 + \left(\frac{\partial y}{\partial \rho}\right)^2 + \left(\frac{\partial z}{\partial \rho}\right)^2 \\
&= \left[\frac{\partial(\rho \cos\theta)}{\partial \rho}\right]^2 + \left[\frac{\partial(\rho \sin\theta)}{\partial \rho}\right]^2 + \left[\frac{\partial z}{\partial \rho}\right]^2 \\
&= \cos^2\theta + \sin^2\theta + 0 = 1,
\end{aligned}
$$

so $h_\rho = 1$.   (8.128)

$$
\begin{aligned}
h_\theta^2 &= \left(\frac{\partial x}{\partial \theta}\right)^2 + \left(\frac{\partial y}{\partial \theta}\right)^2 + \left(\frac{\partial z}{\partial \theta}\right)^2 \\
&= \left[\frac{\partial(\rho \cos\theta)}{\partial \theta}\right]^2 + \left[\frac{\partial(\rho \sin\theta)}{\partial \theta}\right]^2 + \left[\frac{\partial z}{\partial \theta}\right]^2 \\
&= \rho^2 \sin^2\theta + \rho^2 \cos^2\theta + 0 = \rho^2,
\end{aligned}
$$

so $h_\theta = \rho$.   (8.129)

$$
\begin{aligned}
h_z^2 &= \left(\frac{\partial x}{\partial z}\right)^2 + \left(\frac{\partial y}{\partial z}\right)^2 + \left(\frac{\partial z}{\partial z}\right)^2 \\
&= \left[\frac{\partial(\rho \cos\theta)}{\partial z}\right]^2 + \left[\frac{\partial(\rho \sin\theta)}{\partial z}\right]^2 + \left[\frac{\partial z}{\partial z}\right]^2 \\
&= 1, \quad \text{so } h_z = 1.
\end{aligned}
$$
                                                                            (8.130)

6. (a) Write the potential of a dipole **p** (in the z direction) in cylindrical coordinates.

   (b) Find **E** of the dipole by taking the negative gradient in cylindrical coordinates. Show that this equals the result of writing the vector expression for **E** in cylindrical coordinates.

Answer:

(a) The potential due to an electric dipole **p** in the z direction is (with $\mathbf{r} = \boldsymbol{\rho} + z\hat{\mathbf{k}}$)

$$
\phi = \frac{\mathbf{p}\cdot\mathbf{r}}{r^3} = \frac{p\hat{\mathbf{k}}\cdot\mathbf{r}}{r^3} = \frac{pz}{(\rho^2 + z^2)^{\frac{3}{2}}}.
$$
                                                                            (8.131)

(b) The components of the electric field are found by taking the negative gradient of the potential. In cylindrical coordinates, these are

$$
E_\rho = -\partial_\rho\phi = -\partial_\rho\left[\frac{pz}{(\rho^2 + z^2)^{\frac{3}{2}}}\right] = \frac{3pz\rho}{(\rho^2 + z^2)^{\frac{5}{2}}},
$$
                                                                            (8.132)

$$E_\theta \;=\; -\frac{1}{\rho}(\partial_\theta \phi) = -\frac{1}{\rho}\partial_\theta\!\left[\frac{pz}{(\rho^2+z^2)^{\frac{3}{2}}}\right] = 0, \tag{8.133}$$

$$E_z \;=\; -\partial_z \phi = -\partial_z\!\left[\frac{pz}{(\rho^2+z^2)^{\frac{3}{2}}}\right] = \frac{3pz^2}{(\rho^2+z^2)^{\frac{5}{2}}} - \frac{p}{(\rho^2+z^2)^{\frac{3}{2}}}. \tag{8.134}$$

The vector expression for **E** due to an electric dipole is

$$\mathbf{E} = \frac{3(\mathbf{p}\cdot\mathbf{r})\mathbf{r}}{r^5} - \frac{\mathbf{p}}{r^3}. \tag{8.135}$$

The cylindrical components of **E** (with $\mathbf{p} = p\hat{\mathbf{k}}$) are given by

$$E_\rho = \hat{\boldsymbol{\rho}}\cdot\mathbf{E} \;=\; \frac{3[\hat{\boldsymbol{\rho}}\cdot(\rho\hat{\boldsymbol{\rho}}+z\hat{\mathbf{k}})][p\hat{\mathbf{k}}\cdot(\rho\hat{\boldsymbol{\rho}}+z\hat{\mathbf{k}})]}{(\rho^2+z^2)^{\frac{5}{2}}} - \frac{p\hat{\mathbf{k}}\cdot\hat{\boldsymbol{\rho}}}{(\rho^2+z^2)^{\frac{3}{2}}}$$

$$= \frac{3\rho z p}{(\rho^2+z^2)^{\frac{5}{2}}}, \tag{8.136}$$

$$E_\theta = \hat{\boldsymbol{\theta}}\cdot\mathbf{E} \;=\; \frac{3[\hat{\boldsymbol{\theta}}\cdot(\rho\hat{\boldsymbol{\rho}}+z\hat{\mathbf{k}})][p\hat{\mathbf{k}}\cdot(\rho\hat{\boldsymbol{\rho}}+z\hat{\mathbf{k}})]}{(\rho^2+z^2)^{\frac{5}{2}}} - \frac{p\hat{\mathbf{k}}\cdot\hat{\boldsymbol{\theta}}}{(\rho^2+z^2)^{\frac{3}{2}}}$$

$$= 0, \tag{8.137}$$

$$E_z = \hat{\mathbf{k}}\cdot\mathbf{E} \;=\; \frac{3[\hat{\mathbf{k}}\cdot(\rho\hat{\boldsymbol{\rho}}+z\hat{\mathbf{k}})][p\hat{\mathbf{k}}\cdot(\rho\hat{\boldsymbol{\rho}}+z\hat{\mathbf{k}})]}{(\rho^2+z^2)^{\frac{5}{2}}} - \frac{p\hat{\mathbf{k}}\cdot\hat{\mathbf{k}}}{(\rho^2+z^2)^{\frac{3}{2}}}$$

$$= \frac{3z^2 p}{(\rho^2+z^2)^{\frac{5}{2}}} - \frac{p}{(\rho^2+z^2)^{\frac{3}{2}}}. \tag{8.138}$$

These cylindrical components agree with those in Eqs. (8.132)-(8.134).

7. (a) Write the potential of a dipole **p** (in a general direction) in cylindrical coordinates.

   (b) Find **E** of the dipole by taking the negative gradient in cylindrical coordinates. Show that this equals the result of writing the vector expression for **E** in cylindrical coordinates.

Answer:

(a) The potential due to an electric dipole **p** is

$$\phi = \frac{\mathbf{p}\cdot\mathbf{r}}{r^3}. \tag{8.139}$$

The vector **r** in this equation is not the radial coordinate vector in cylindrical coordinates. Because of this difference, we will use the symbol $\boldsymbol{\rho}$ for the radial coordinate vector. Then the vector **r** is given by $\mathbf{r} = \rho\hat{\boldsymbol{\rho}} + z\hat{\mathbf{k}}$, and $\mathbf{p} = p_\rho\hat{\boldsymbol{\rho}} + p_z\hat{\mathbf{k}}$. The dipole potential is given in cylindrical coordinates by

$$\phi = \frac{\mathbf{p}\cdot\mathbf{r}}{r^3} = \frac{\rho p_\rho + z p_z}{(\rho^2 + z^2)^{\frac{3}{2}}}. \tag{8.140}$$

(b) The components of the electric field are found by taking the negative gradient of the potential. In cylindrical coordinates, these are

$$E_\rho = -\partial_\rho\phi = -\partial_\rho\left[\frac{\rho p_\rho + z p_z}{(\rho^2 + z^2)^{\frac{3}{2}}}\right] = \frac{3\rho(\rho p_\rho + z p_z)}{(\rho^2 + z^2)^{\frac{5}{2}}} - \frac{p_\rho}{(\rho^2 + z^2)^{\frac{3}{2}}}, \tag{8.141}$$

$$E_\theta = -\frac{1}{\rho}(\partial_\theta\phi) = -\frac{1}{\rho}\partial_\theta\left[\frac{\rho p_\rho + z p_z}{(\rho^2 + z^2)^{\frac{3}{2}}}\right] = 0, \tag{8.142}$$

$$E_z = -\partial_z\phi = -\partial_z\left[\frac{\rho p_\rho + z p_z}{(\rho^2 + z^2)^{\frac{3}{2}}}\right] = \frac{3z(\rho p_\rho + z p_z)}{(\rho^2 + z^2)^{\frac{5}{2}}} - \frac{p_z}{(\rho^2 + z^2)^{\frac{3}{2}}}. \tag{8.143}$$

The vector expression for $\mathbf{E}$ due to an electric dipole is

$$\mathbf{E} = \frac{3(\mathbf{p}\cdot\mathbf{r})\mathbf{r}}{r^5} - \frac{\mathbf{p}}{r^3}. \tag{8.144}$$

The cylindrical components of $\mathbf{E}$ are given by

$$E_\rho = \hat{\boldsymbol{\rho}}\cdot\mathbf{E} = \frac{3[\hat{\boldsymbol{\rho}}\cdot(\rho\hat{\boldsymbol{\rho}} + z\hat{\mathbf{k}})][\mathbf{p}\cdot(\rho\hat{\boldsymbol{\rho}} + z\hat{\mathbf{k}})]}{(\rho^2 + z^2)^{\frac{5}{2}}} - \frac{\hat{\boldsymbol{\rho}}\cdot\mathbf{p}}{(\rho^2 + z^2)^{\frac{3}{2}}}$$

$$= \frac{3\rho(\rho p_\rho + z p_z)}{(\rho^2 + z^2)^{\frac{5}{2}}} - \frac{p_\rho}{(\rho^2 + z^2)^{\frac{3}{2}}}, \tag{8.145}$$

$$E_\theta = \hat{\boldsymbol{\theta}}\cdot\mathbf{E} = \frac{3[\hat{\boldsymbol{\theta}}\cdot(\rho\hat{\boldsymbol{\rho}} + z\hat{\mathbf{k}})][\mathbf{p}\cdot(\rho\hat{\boldsymbol{\rho}} + z\hat{\mathbf{k}})]}{(\rho^2 + z^2)^{\frac{5}{2}}} - \frac{\hat{\boldsymbol{\theta}}\cdot\mathbf{p}}{(\rho^2 + z^2)^{\frac{3}{2}}}$$

$$= 0, \tag{8.146}$$

$$E_z = \hat{\mathbf{k}}\cdot\mathbf{E} = \frac{3[\hat{\mathbf{k}}\cdot(\rho\hat{\boldsymbol{\rho}} + z\hat{\mathbf{k}})][\mathbf{p}\cdot(\rho\hat{\boldsymbol{\rho}} + z\hat{\mathbf{k}})]}{(\rho^2 + z^2)^{\frac{5}{2}}} - \frac{\hat{\mathbf{k}}\cdot\mathbf{p}}{(\rho^2 + z^2)^{\frac{3}{2}}}$$

$$= \frac{3z(\rho p_\rho + z p_z)}{(\rho^2 + z^2)^{\frac{5}{2}}} - \frac{p_z}{(\rho^2 + z^2)^{\frac{3}{2}}}. \tag{8.147}$$

These cylindrical components agree with those in Eqs. (8.141)-(8.143).

8.7   Dyadics

1. Find the dipole moment of

(a) a straight wire of length L with a linear charge density
$\lambda(z) = \lambda_0 z/L$, $|z| < L/2$.

(b) a hollow sphere of radius R with a surface charge distribution
$\sigma(\theta) = (q/R^2)\cos(\theta)$.

Answer:

(a) The electric dipole moment of a straight wire of length L with a linear
charge density $\lambda(z) = \lambda_0 z/L$, $|z| < L/2$ is given by

$$p = \int_{-L/2}^{+L/2} \lambda z dz = \frac{\lambda_0}{L}\int_{-L/2}^{+L/2} z^2 dz = \frac{\lambda_0 L^2}{12}. \tag{8.148}$$

(b) The dipole moment of a hollow sphere of radius R with a surface
charge distribution $\sigma(\theta) = (q/R^2)\cos\theta$ will be in the z direction and
is given by

$$p_z = \oint z\sigma dA = \oint R^3 d\Omega(q/R^2)\cos^2\theta$$

$$= 2\pi q R \int_0^\pi \cos^2\theta \sin\theta d\theta = \frac{4\pi}{3} q R. \tag{8.149}$$

2. Calculate the quadrupole moment for

(a) a uniformly charged needle of length **L** and charge q.

(b) a uniformly charged disk of radius R and charge q.

(c) a spherical shell of radius R with surface charge distribution $(q/R^2)\cos^2\theta$.

Answer:

(a) The quadrupole moment of a uniformly charged needle of length L
and charge q is

$$Q_0 = \frac{1}{2}\int_{-L/2}^{+L/2} (2z^2 - x^2 - y^2)(q/L)dz$$

$$= \frac{1}{2}\frac{q}{L}\int_{-L/2}^{+L/2} 2z^2 dz = \frac{1}{12} q L^2, \tag{8.150}$$

where q/L is the charge per unit length, and $x = 0$, $y = 0$ for a
straight wire on the z axis.

(b) The quadrupole moment of a uniformly charged disk of radius R and charge q is

$$
\begin{aligned}
Q_0 &= \frac{1}{2} \int (3\cos^2\theta - 1) r^2 \sigma dA \\
&= \frac{1}{2} \int_0^R (3\cos^2\theta - 1) r^2 (q/\pi R^2) 2\pi r dr \\
&= -\frac{q}{R^2} \int_0^R r^3 dr = -\frac{1}{4} q R^2, \qquad (8.151)
\end{aligned}
$$

where $\sigma = (q/\pi R^2)$ is the surface charge density, and $\cos^2\theta = 0$ for a disk whose axis is aligned along the z axis.

(c) The quadrupole moment of a spherical shell of radius R with a surface charge distribution $\sigma = (q/4\pi R^2)\cos^2\theta$ is

$$
\begin{aligned}
Q_0 &= \frac{1}{2} \int (3\cos^2\theta - 1) r^2 \sigma dA \\
&= \frac{1}{2} \int_0^\pi (3\cos^2\theta - 1) R^2 \left(\frac{q}{4\pi R^2}\right) \cos^2\theta 2\pi R^2 \sin\theta d\theta \\
&= \frac{q R^2}{4} \int_0^\pi (3\cos^4\theta - \cos^2\theta) \sin\theta d\theta = \frac{2}{15} q R^2. \qquad (8.152)
\end{aligned}
$$

3. Calculate the quadrupole moment for

   (a) two point charges, each of charge $+q$, a distance L apart with a third collinear point charge, -2q, at their midpoint.

   (b) a point charge, $+q$, at the center of a circular uniform line charge, -q, at radius L.

Answer:

(a) The quadrupole moment of two point charges, $+q$, a distance L (along the z axis) apart, with a third collinear point charge, -2q, at their midpoint is given in Cartesian coordinates by

$$
\begin{aligned}
Q_0 &= \frac{1}{2} \sum_i (2z_i^2 - x_i^2 - y_i^2) q_i, \quad \text{for point charges} \\
&= q(L/2)^2 + q(-L/2)^2 + 0 \\
&= q L^2/2. \qquad (8.153)
\end{aligned}
$$

(b) The quadrupole moment of a point charge, $+q$, at the center of a circular uniform line charge, -q, at radius L is given in spherical coordinates (with the axis of the circular loop along the z axis) by

$$
Q_0 = \frac{1}{2} \int_0^{2\pi} (3\cos^2\theta - 1) L^2 (q/2\pi L) L d\phi = -\frac{1}{2} q L^2, \qquad (8.154)
$$

since every point on the circular ring has $\cos\theta = 0$, and the charge
at the origin does not contribute to Q_0.

4. (a) Show, using the quadrupole dyadic $[\mathbf{Q}]$, that the potential for a symmetric quadrupole is given by

$$\phi = \frac{Q_0}{2r^3}[3(\hat{\mathbf{k}}\cdot\hat{\mathbf{r}})^2 - 1]. \tag{8.155}$$

(b) Show that the electric field, given by the negative gradient of this potential, is

$$\mathbf{E} = \frac{3Q_0}{2r^4}[5(\hat{\mathbf{k}}\cdot\hat{\mathbf{r}})^2\hat{\mathbf{r}} - 2(\hat{\mathbf{k}}\cdot\hat{\mathbf{r}})\hat{\mathbf{k}} - \hat{\mathbf{r}}]. \tag{8.156}$$

Answer:

(a) The quadrupole dyadic for a symmetric quadrupole is

$$[\mathbf{Q}] = \frac{1}{2}Q_0[3\hat{\mathbf{k}}\hat{\mathbf{k}} - \hat{\mathbf{n}}]. \tag{8.157}$$

The quadrupole potential is given by

$$\begin{aligned}
\phi(\mathbf{r}) &= \frac{\hat{\mathbf{r}}\hat{\mathbf{r}} : [\mathbf{Q}]}{r^3} \\
&= \frac{\hat{\mathbf{r}}\hat{\mathbf{r}} : Q_0[3\hat{\mathbf{k}}\hat{\mathbf{k}} - \hat{\mathbf{n}}]}{2r^3} \\
&= \frac{Q_0}{2r^3}[3(\hat{\mathbf{k}}\cdot\hat{\mathbf{r}})^2 - 1]. \tag{8.158}
\end{aligned}$$

(b) The quadrupole field is given by

$$\begin{aligned}
\mathbf{E} &= -\nabla\phi = -Q_0\nabla\left[\frac{3(\hat{\mathbf{k}}\cdot\hat{\mathbf{r}})^2 - 1}{2r^3}\right] \\
&= -Q_0\nabla\left[\frac{3(\hat{\mathbf{k}}\cdot\mathbf{r})^2}{2r^5} - \frac{1}{2r^3}\right] \\
&= -Q_0\left[\frac{-15(\hat{\mathbf{k}}\cdot\mathbf{r})^2\hat{\mathbf{r}}}{2r^6} + \frac{6\hat{\mathbf{k}}(\hat{\mathbf{k}}\cdot\mathbf{r})}{2r^5} + \frac{3\hat{\mathbf{r}}}{2r^4}\right] \\
&= \frac{3Q_0}{2r^4}[5(\hat{\mathbf{k}}\cdot\hat{\mathbf{r}})^2\hat{\mathbf{r}} - 2(\hat{\mathbf{k}}\cdot\hat{\mathbf{r}})\hat{\mathbf{k}} - \hat{\mathbf{r}}]. \tag{8.159}
\end{aligned}$$

5. A plywood board of mass M has dimensions $6a$ by $8a$, and a negligible thickness.

(a) Find its principal moments of inertia.

(b) The board rotates with angular velocity $\boldsymbol{\omega}$ about an axis along one of its diagonals. Find its angular momentum.

Answer:

(a)

$$I_x = \frac{M}{48a^2} \int_{-3a}^{3a} dx \int_{-4a}^{4a} y^2 dy = \left(\frac{16}{3}\right) Ma^2, \qquad (8.160)$$

$$I_y = \frac{M}{48a^2} \int_{-3a}^{3a} x^2 dx \int_{-4a}^{4a} dy = 3Ma^2, \qquad (8.161)$$

$$I_z = \frac{M}{48a^2} \int_{-3a}^{3a} dx \int_{-4a}^{4a} dy (x^2 + y^2)$$

$$= I_x + I_y = \left(\frac{25}{3}\right) Ma^2. \qquad (8.162)$$

(b)

$$\begin{aligned}
\mathbf{L} &= [\mathbf{I}]{\cdot}\boldsymbol{\omega} = I_x \omega_x + I_y \omega_y \\
&= Ma^2\omega \left[\left(\frac{16}{3}\right)\hat{\mathbf{i}}\cos\theta + 3\hat{\mathbf{j}}\sin\theta\right], \qquad \tan\theta = \frac{3}{4} \\
&= Ma^2\omega \left[\left(\frac{16}{3}\right)\left(\frac{3}{5}\right)\hat{\mathbf{i}} + 3\left(\frac{4}{5}\right)\hat{\mathbf{j}}\right] \\
&= \frac{Ma^2\omega}{5}(16\hat{\mathbf{i}} + 12\hat{\mathbf{j}}). \qquad (8.163)
\end{aligned}$$

Index